GETTING STARTED IN HEATING AND AIR CONDITIONING SERVICE

Third Edition

Allen Russell

Business News Publishing Company • Troy, Michigan

Library of Congress Cataloging in Publication Data

Russell, Allen.
   Getting started in heating and air conditioning service.
   Includes index.
   1. Furnaces — Maintenance and repair. 2. Air conditioning — Equipment and supplies — Maintenance and
repair. I. Title.
TH7623.R87 1977   697   77-13571
ISBN 0-912524-17-0

Printed in the United States of America

To the Reader

This book on heating and air conditioning is a new approach to the age-old problem of preparing younger people for the world of work. You will not be required to memorize theories and formulas or waste time on *exercises* which have little, if any, relationship to your future employment. Rather than giving you a broad theoretical background, this book concentrates on practical, *how to do* procedures that you must master to become a well-trained specialist in heating and air conditioning service. The book should become your daily reference, both while you are learning and after you become gainfully employed by a warm air heating and air conditioning contractor.

You will note that heating is allotted equal time with the air conditioning because the two depend completely upon each other. Appliances, commercial cases, window coolers, etc. are not included because they belong to a different industry. This is an age of specialists and, with the proper guidance and motivation, you will become just that --- a specialist in heating and air conditioning service with a re-warding career ahead of you.

This book will give you all of the basic heating and cooling fundamentals because, in order to do *any* service work properly, the serviceman must thoroughly understand what the equipment or component is *supposed* to do. Then he can determine whether or not it is malfunctioning and follow a logical sequence to locate and correct the problem. In addition to the fundamentals, this book will also give basic maintenance information, show the proper use of common service tools and give you the skills needed to perform minor service.

This book is written for the beginner and covers all the basics required for more advanced work. It can also serve as a refresher for men already in the business, since many new products, not shown in other works, are covered.

This book covers a broad cross section of the products of many manufacturers to familiarize the serviceman with the variety of equipment he will find in the field. It has been written with the cooperation and assistance of many manufacturers in the heating and air conditioning industry and to each of them we extend our thanks for their help.

Those who wrote and contributed to this book have had many years of experience at all levels of this industry and know its needs well. They have observed and participated in the training activities of major manufacturers of heating and air conditioning equipment and, from this experience, have designed this book for you. Study it closely, learn it well and your future will grow with the industry.

Contents

Introduction

Before proceeding, it might be advisable to examine the heating and air conditioning industry to get an idea of what it's all about. New names, such as the *Human Comfort Industry* or *Environmental Control Industry* have replaced the original terms of *refrigeration* or *refrigeration and air conditioning*. This is a realistic trend since it indicates the much broader scope of the industry as it exists today. Heating systems, which often serve as the prime air mover for cooling, have been given more prominence because the well-rounded serviceman, who is most in demand today, must know both heating and air conditioning.

Because man is a constant temperature animal, his body temperature must remain close to 98.6° F in order for him to survive. Man is said to have originated in the tropic zones of Africa where keeping warm was never a problem. However, as he ranged out of the tropics and into colder climates, keeping warm became a necessity. His early discovery of fire, plus the use of animal skins for cover, allowed man to range over all the areas of the world, even the polar regions where temperatures drop to 100° F below zero.

All types of fuels have been burned to provide warmth, from *buffalo chips* (dung) to wood and natural gas. The early Romans even designed hot water radiant heat systems not too unlike those used today. Fire is used for heat and comfort in all parts of the world and the development of today's furnace, regardless of fuel or energy source, is a natural outgrowth of man's search for more efficient and convenient comfort. Today's furnaces give accurately controlled, dependable, even heat and they may be installed simply and economically. They are small in size and attractive in appearance.

In temperate zones such as the United States, man had to adjust not only to cold weather but also to extremely hot weather. Once the problem of keeping warm was solved, he then turned his attention to summertime comfort.

Comfort cooling, with dependable, economical equipment, is a fairly recent development whose beginnings date back only to the late 1920's. The cooling or air conditioning industry really didn't take hold and command a wide market until the late 1940's.

The intent of this book is to prepare the student to install, maintain and service residential furnaces (including gas, oil and electric energy sources) and electric air conditioning, and to serve as an introduction to some specialized products such as mobile home units, heat pumps and the like.

The January 29, 1979 Statistical Issue of the industry trade paper Air Conditioning, Heating & Refrigeration News reported the following factory sales in 1978.

	1978 Sales
Unitary Air Conditioning*	2,675,000
Gas Furnaces	1,619,000
Oil Furnaces	228,000
Electric Furnaces	350,000
Electronic Air Cleaners	210,000
Humidifiers	600,000
Heat Pumps (air to air)	560,000

Includes heat pumps, packaged terminal air conditioners and rooftop units. 95% are rated at 5 tons and under.

According to the 1976 Housing Census, there were 79,316,000 all-year housing units. Over 40,470,000 units were equipped with central or room air conditioners and 40,720,000 were heated with warm air furnaces. Electric heating is in use in another 5,217,000 units.

Over 2,019,000 new homes were started in 1978, of which approximately 95% had central heating. Of these, over ¾ were ducted systems. Furthermore, 54% included central air conditioning. In addition, nearly 275,600 mobile homes were produced in 1978, most of them with a ducted heating system and the potential for add-on air conditioning.

Just as important is the current emphasis on energy conservation. A good serviceman can adjust an existing heating system to save energy as well as cost of operation.

This, therefore, is a large and rapidly expanding industry with a great demand for trained manpower. A good question to ask is, "What kind of a man is the modern service specialist and how much knowledge is required to really move ahead in the business?"

The first requirement is an average degree of basic intelligence, the second is a liking for mechanical/electrical equipment. There is very little electronic equipment used in this industry, but wiring, electric circuitry, motors and other mechanical devices are used in all types of equipment. Some manufacturers and many schools use standard mechanical aptitude tests to predetermine a student's ability to master the required skills.

The service specialist should also know something about the related trades: carpentry to understand house construction, sheet metal to know duct systems, air movement to understand comfort, welding and brazing to understand piping, and so forth.

A service specialist today is not a *grease monkey*. Almost all basements are clean and the equipment does not normally build up dirt, sludge or oil to any great degree. More and more basements are being turned into recreation rooms and the equipment has been designed to be compatible with this trend. The serviceman frequently drives a new, clean service truck, his tools and instruments are supplied by his employer and he is the *front line* contact with the customer. Age varies a great deal, but typically he is between 20 and 28. He has a very necessary skill and is in demand everywhere.

For many reasons, certain products are more popular in some parts of the country than others, so he must be well-rounded and flexible. For instance, oil equipment is almost nonexistent in Texas but it is extremely popular in North Carolina and New England. Electric heat, particularly heat pumps, is very popular in Tennessee, Florida and the Pacific Northwest where electric rates are low and the climate is suitable for this type of application, but limited in the northern tier of states. Therefore the skills required will vary according to your region.

In March 1973, the industry association, Air-Conditioning & Refrigeration Institute (ARI), updated its survey of manpower needs, originally published in 1968 and now projected to 1980. The original survey was based on the manpower shortage which then existed and did not take into account replacements needed due to death or retirements. It also assumed an industry growth rate, adjusted for inflation, of 10% per year when, in fact, the industry has grown 12 to 13% per year during this period. Their 1968 estimate for all positions was 64,000 *new*

employees by 1975 and an additional 77,000 by 1980. In view of the known expansion of the industry since 1973, a figure of 100,000 new men might be a closer estimate now. By far the greatest number of these will be servicemen.

The types of organizations hiring men with these skills are as follows:

Warm air heating and air conditioning dealers
Electrical contractors
Sheet metal contractors
Mechanical contractors
Independent service companies
Utilities
Wholesalers
Distributors
Manufacturing companies
Mass merchandisers
Fuel suppliers, oil and LP gas companies
Fast-food chains

All of these organizations have some service responsibility, ranging from the constant and continuing responsibility of the heating and air conditioning contractor, to the one-year responsibility of an installing mechanical contractor.

Pay scales will vary depending upon the region, the size of dealer, whether the serviceman is union or nonunion and upon the actual skill possessed by the individual. Overall, pay scales can range from $3 per hour up to $9 or $10 per hour, plus a fair amount of overtime.

There is a good prospect for job security because *consumerism* is here to stay and contractors, to a greater extent, are being held responsible and *liable* for the equipment they install. Hence, they must have qualified men on their payroll at all

times. This used to be a seasonal business, but this is no longer true. There is a break between the heating and air conditioning seasons, but these gaps have been filled by the very rapid growth of preventive maintenance, making heating and air conditioning service a year-around occupation. This year-around demand almost assures that the dealer will make money on service, something he was never able to do before, primarily because he couldn't keep his men busy the year-around.

Service of a furnace or air conditioner cannot be postponed by the homeowner. If the furnace goes out when it's minus 4° F, he doesn't wait a month to get it repaired. The same is true if an air conditioning system fails when the outside temperature is 100° F. Therefore, the serviceman's availability is important and necessary to the owner of a service-related business. There are just so many trained servicemen available today. So, until many, many more are trained, their skills will continue to be in demand. With the continued growth of the industry, it appears that there will be a shortage of trained servicemen for some time to come.

Recognition of this occupation, as a special skill, has now been approved by the Department of Labor's Bureau of Apprenticeship. Aid programs for On-the-Job Training, under the Veterans Administration, and school co-op programs are available in many areas, so financial help is available.

There is an opportunity for a rewarding career in this field, the pay is good and the working conditions are great. Now, let us concentrate on what is needed to be successful.

Chapter 1
BASIC ELECTRICITY

This is an introduction to electricity, including types of circuits, wiring, fuses, switches, drawing pictorial and schematic circuits, controlling flow of current and understanding what is required to electrically operate and control furnaces and air conditioners.

Safety should be of prime consideration at all times, since uncontrolled electrical circuits or careless handling can be harmful and dangerous. *When dealing with any "live" circuit, particularly line voltages of 120v or higher, use extreme caution and proceed carefully at all times.* Treat all circuits as *live* until you have checked and positively determined that they are not. Extreme caution should be exercised if there is any standing water in the area from a condensate drain, humidifier or any other source. Standing in water *grounds* the body and touching a live wire allows current to pass through the body to ground—*this can be fatal.*

ELECTRON THEORY

Much is known about how electricity acts, what it does, and its basic makeup. However, its exact nature is still a mystery. Electricity is a form of energy and, as such, can neither be created nor destroyed. However, it can be transformed into other forms of energy such as heat and light and can do work when applied to an electric motor.

To understand how this is possible, you must understand the makeup of matter. *Matter* is any substance that occupies space and has weight. If it were possible to break down a substance into smaller and smaller units, eventually a very tiny piece

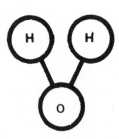

FIG. 1-1.   A water molecule is composed of two atoms of hydrogen and one atom of oxygen.

FIG. 1-2.   An atom consists of a number of electrons in orbit around a nucleus containing an equal number of protons.

would be left. This piece—too small to be seen with the most powerful microscope—is called a molecule, Figure 1-1. Molecules are made up of even tinier pieces called atoms. An atom is defined as the smallest part of a substance still retaining the original characteristics of the substance, of which it is composed, Figure 1-2. There are millions of atoms in the smallest piece of the substance.

The center, or nucleus, of an atom is made up of smaller particles called protons and neutrons, and surrounded by a quantity of other particles, spinning around the center, called electrons. Electrons and their movement are the key to how electricity acts and current flows. Different materials have different combinations of electrons and protons which give them different electrical characteristics. This will be covered later.

Experiments have shown that electrons have a negative charge, protons have a positive charge and neutrons are neither plus nor minus. Further, opposite charges are attracted to each other and like charges repel each other, Figure 1-3. In other words, an electron (- charge) is attracted to a proton (+ charge), but electrons repel each other. The protons do not move, but form the center of the atom with the neutrons. The electrons move rapidly, in orbit, around the nucleus or center and are attracted to the protons because of different charges (+ and -). The centrifugal force, created by their spinning, offsets this attraction and keeps the electrons in place around the nucleus, Figure 1-4. Centrifugal force can be demonstrated by whirling around an object

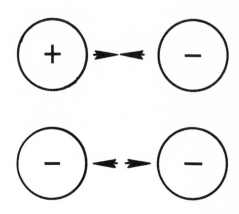

FIG. 1-3.   Opposite charges attract, like charges repel.

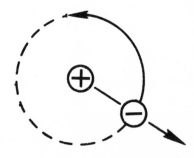

FIG. 1-4.   The spinning motion of the electron offsets the attraction of the oppositely-charged proton.

attached to a string. If the string is released, the object flies outward rather than dropping straight down.

Since the atoms tend to remain stable, with a constant number of electrons and protons, an additional force is necessary to move the electrons from one atom to another and create electrical flow. Electrons will move from minus to plus, or from negative to positive. If a deficiency of electrons is created (that is, the atom has fewer electrons than it is supposed to have), then the electron balance is unequal and a *potential difference of electric force* will exist. This is also called electromotive force (emf) or voltage, which will act to push the electrons from one atom to another, resulting in electron flow, Figure 1-5. Some atoms have extra or free electrons and these are most easily moved from one atom to the next by the emf.

Electrons always travel at a speed of 186,000 miles per second or the same as the speed of light. This will be a constant rate, but the number or density of the electrons passing a given point at a given instant can change, and thus greater or smaller current flow will result. This current flow is measured in *amperes*.

All materials will resist the flow of electrons to a certain extent. The amount of this resistance to flow, which is measured in *ohms*, determines whether the substance is considered a *conductor* or an *insulator*. Glass, plastics, and mica have a high resistance to electron flow and are used extensively as insulators, Figure 1-6. Many metals such as copper, silver, aluminum and platinum have an excess of electrons and so pass electrons freely. These

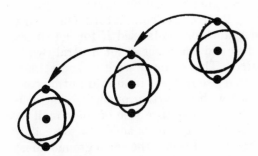

FIG. 1-5 Emf pushes electrons from one atom to another, creating electron flow or current.

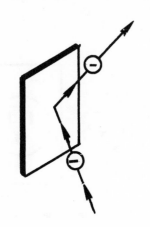

FIG. 1-6. Insulators resist electron flow.

FIG. 1-7. Heat increases the resistance to electron flow.

are normally used as conductors since it takes relatively little force to start and maintain electron flow.

Even with good conductors, some other factors will increase their resistance to electron flow. Outside or ambient temperature (the temperature around the wire) is an important factor. Very little resistance is offered when the surrounding air is cold. However, if the wire is confined in a small area where the temperature can build up or if operating in hot surrounding atmosphere, resistance to current flow increases, Figure 1-7. Wires are insulated not only from each other, but also to reduce the effect of ambient temperature.

Similarly, the area or size of the wire can affect flow. If the cross-sectional area of the wire is too small for the current carried, it will warm up due to friction and reduce the flow of electrons. Excessive heat can also build up and burn the insulation, causing a short. There is no problem if the wire is oversized as this reduces friction, which reduces the amount of heat generated.

SUMMARY

Negatively charged electrical particles, called electrons, flow through a material when an electromotive force is applied, causing them to move from atom to atom. This *emf* is measured in *volts*. The movement of the electrons is called *current* and the amount of current or density of the electrons is measured in *amperes*. In any material, there is some resistance to current flow which will vary with the material itself, the surrounding temperature, and the size of the conductor. *Resistance* is measured in *ohms*.

ALTERNATING CURRENT (AC)

The most common voltage encountered is 120v, single phase, 60 cycle (or hertz as it is now called), alternating current. It is called alternating current because the direction of current flow changes 60 times a second, flowing first in one direction and then the other. This characteristic of alternating current is due to the way it is generated by the power company. The pattern can be seen in a diagram of one complete cycle, Figure 1-8. Starting from

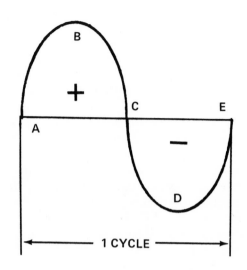

FIG. 1-8.   Alternating current that changes its direction and polarity 60 times a second is called 60-cycle ac.

zero, at point A, the voltage builds to a positive peak 90° from A at point B, and then diminishes to zero again at point C. It then builds to a maximum minus value at point D and returns to zero at point E. Thus, during a cycle, half the time current flows in one direction and the other half of the time it flows in the opposite direction.

Single phase, 120v current requires only 2 wires for conductors. On wiring diagrams, one is labeled *H* for hot (black wire) and the other *N* for neutral (white wire). The neutral wire is sometimes called ground but it should not be confused with an earth ground. It is the neutral or zero voltage side of the circuit. A third wire (green) is added to some 120v circuits and this is the earth ground. Most new installations include this ground wire, which requires a three prong plug outlet, Figure 1-9. The ground is not normally shown on wiring diagrams as the third wire.

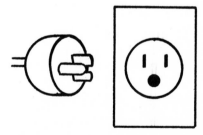

FIG. 1-9.   Three-prong outlet assures positive grounding.

REQUIREMENTS FOR A CIRCUIT

For electricity to do work, three things must be provided: First, there must be a source of emf which is normally supplied by the power company, but it could come from a battery or a generator. Then, there must be resistance to current flow (load) so the electrical energy will be transformed into work, heat, light or some other form of energy. Third, there must be a continuous path of conductors from the source to the load and then back to the source, Figure 1-10.

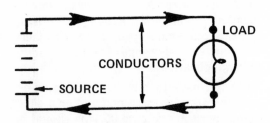

FIG. 1-10.   A typical circuit with a source, load and conductors.

FUSES

A circuit may have a single fuse or it may have a main fuse and several branch circuit fuses. All 120v circuits must have at least one fuse in the *H* or hot leg and it should be located as close as possible to the entrance point. The fuse protects the wiring and other components of the circuit from burning up due to overcurrent or shorts. There are many types of fuses, Figure 1-11: some screw into a socket, others clamp into a bracket, but all perform the same function.

The terms *overcurrent* and *short* are sometimes confused. They are *not* the same thing. Fuses do protect against both, but the timing is different. For example, overcurrent occurs when an added appliance is put on a fully loaded line that was designed and fused for 15 amps, making it draw over 15 amps, Figure 1-12. The higher current draw will overheat the wire, but the *rate* of increased heat will depend on the additional amperes (amps) drawn. If the amperage on the 15 amp circuit is only 17 amps, it may take several minutes for the fuse to blow, or it may not blow at all if the load is reduced in a short period of time. A short, on the other hand, overheats the wire to a far greater degree and the fuse must blow *immediately*.

In the case of a direct short, the current will bypass the load and the amount of electron flow will increase instantaneously, due to lack of resistance. The increased flow of electrons will significantly increase the friction in the wire and it will heat up very rapidly, Figure 1-13. It is therefore necessary to stop the flow of electrons before the insulation on the wire

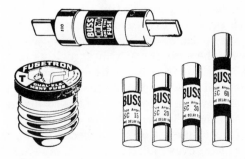

FIG. 1-11.   Examples of the many types of fuses.

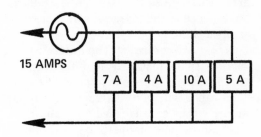

FIG. 1-12.   Four-outlet circuit may be loaded with a TV (7 amps), lights (4 amps), iron (10 amps), or vacuum cleaner (5 amps). The fuse will blow if the iron and any other two appliances are on together. However, the fuse may not blow if only the iron and TV are on together.

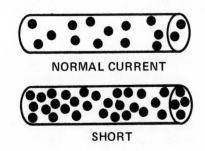

FIG. 1-13.   A shorted wire heats up rapidly.

burns up and starts a fire.

On *overcurrent*, the element in the fuse is also in the line and heats up along with the wire. It has a lower melting point than the wire, so it will melt, stopping current flow before the insulation on the wire burns. It does this in the following sequence, Figure 1-14: The center portion of the element melts first, because it has a smaller cross-sectional area, making it heat up faster than the remainder of the element. Because the gap is small, the circuit is not immediately broken, instead an arc occurs across the gap. This arc burns the balance of the material away, creating an open circuit.

On a short circuit, the entire center section heats instantly and the extreme temperature causes it to turn to vapor. This vapor will carry current and support an arc so the case is filled with an arc-extinguishing noncombustible gas that cools and condenses the vapor. This stops the arc and breaks the circuit.

A number of different types of fuses are available for various circuit applications. The simple strip fuse, either in a screw-in housing or bar-type, is fine for circuits with only resistance loads such as lights, electric heater, etc. They come in many amperage ratings and should be sized to the circuit load.

Motors in a circuit present a different problem because, on start-up, they will draw 5 to 7 times the amperage required for running. Using a fuse large enough for the starting load will protect for starting current and shorts, but will be too large to give any protection against lesser overcurrent while running. If a smaller fuse is

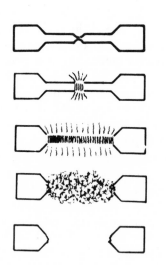

FIG. 1-14.   On overcurrent, a fuse element melts, arcs, vaporizes, and finally opens.

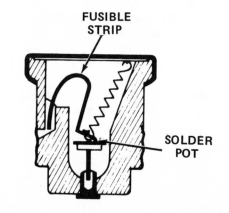

FUSIBLE STRIP

SOLDER POT

FIG. 1-15.   A time delay fuse has a fusible strip to protect against shorts and a solder pot to protect against persistent overcurrent.

used, it will create a nuisance by blowing on start-up and thus prevent the motor from ever starting.

The solution to this is a dual-element fuse called a *time delay* fuse, Figure 1-15. One element is a metal strip that is designed to carry much higher currents for a short time, yet still protect against a direct short. It has been pointed out that, on overcurrent, there is a time lag between the application of additional current and the time the fuse blows. Since a motor, operating normally, will come up to speed rather quickly (less than a second), this short delay before blowing the fuse gives adequate protection. For overload protection, there is a second element, a solder pot. If current overload continues beyond the time required for the motor to come up to speed, it will melt the solder and release a spring that snaps open the circuit. These fuses have a wide range of ratings to serve many applications.

Any time a fuse blows, there is an overload (excessive current flow) or a short somewhere in the circuit. This should be located and corrected before reapplying power. When replacing any type of fuse, do not insert the fuse in a live circuit— open the disconnect first. Inserting a fuse into a live circuit will easily cause an arc which produces a burr on the fuse cap or blade. This burr will prevent good contact. Inspect the insides of the fuse holders. If they are not bright and clean, use an emery cloth to clean them up. Also make sure that the fuse makes good contact with the fuse holder. Figure 1-16 shows only about 20% contact being made, Thus, the fuse will only carry about 20% of its rated amperage and any additional current will cause overheating. This can be checked by attempting to insert a narrow piece of thin paper between the clips and the fuse cap or blade. If the paper can be inserted any distance at all, it indicates that good contact is not being made and the clips should be bent inward to give a tighter contact, Figure 1-17.

Bussman Manufacturing Company, in their booklet *Fuseology* states that whenever a fuse blows, it is due to:

A Short Circuit Be sure that you have checked thoroughly to determine what caused the short circuit and have it corrected before inserting a new fuse.

Overloaded Circuit Check the total circuit amperage draw and remove at least one item in the circuit before inserting the new fuse. Put the item that you have removed back in the circuit. If the fuse blows again, you know that you have an overloaded circuit. This will have to be corrected before an additional fuse is inserted.

Poor Contact Poor contact can occur because of vibration or ambient heat or bending the fuse holder out of shape on the initial insertion. Poor contact is frequently indicated by discoloration of the contact surfaces, on either the fuse holder or the fuse itself. The fuse holder is a contact-making device and must hold the fuse firmly and in the proper position to assure positive contact. Both sides of a spring-type clip must be in full contact with the fuse. If the clip is bent out of shape, Figure 1-18, the fuse holder should be reformed to make proper contact. Any loose connection or improper contact will gener-

FIG. 1-16.   Poor contact with the fuse holder can reduce fuse capacity by 80%.

FIG. 1-17.   Using a thin strip of paper to check fuse contact with the holder.

FIG. 1-18.   Bent fuse holders should be reformed to assure maximum contact.

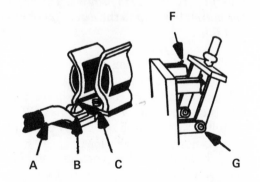

FIG. 1-19.   All connections to fuse holders and switches should be secure.

ate increased heat and can cause the fuse to blow. In Figure 1-19, lug *A* must be properly soldered to the wire or cable, bolt *B*, holding lug to the fuse holder, must be tightly drawn up and screw *C*, holding the clip to fuse holder, must hold the clip firmly. Switch blade clips *F* and switch hinge *G* must be tight.

Wrong Size Fuse The proper size fuse must be used, following manufacturer's instructions. If the fuse is too small, it will immediately blow because it cannot handle the load on the circuit. If the fuse is too large, it will not protect the circuit and will only blow when a serious problem occurs. Usually, this will be too late.

High Ambient Temperature If the air around the fuse is very hot, for instance, in an enclosed box with the heater or some other source of heat, it will have a tendency to blow. Maximum temperature is approximately 125° F and anything beyond this will result in inadequate protection.

Vibration Vibration can cause the fuse holder to change its shape and not make proper contact with the fuse. When this happens, additional heat is generated and the fuse may blow, even before the current becomes excessive.

Fuses are designed to give *one-time* protection and must be replaced if they blow. Another type of circuit protection has a manual reset feature if an overload occurs. This is called a *circuit breaker* and is most often used as the main protection at the service entrance for all branch circuits. See Figure 1-20. It operates on the same principle as a fuse in that it senses heat from overcurrent and trips a switch when the heat is excessive. Some small fuses are available which are manual re-set, and these are usually used on controls or instrument panels.

FIG. 1-20.   A circuit breaker can be reset after tripping due to overcurrent. *Courtesy Square D Co.*

SWITCHES

The operation of all electrical devices in the circuit is controlled by some type of switch located in the *hot* line, ahead of the devices being controlled. Switches are first classified by the number of *poles*. A *single pole* switch has one hot line, or source, a *double pole* switch has two hot lines. Switches are next classified by the number of *throws*. A simple on-off switch, controlling one load, is a *single throw* switch. A switch controlling alternate loads is a *double throw* switch. An *off* position is not considered a load since no power is consumed. One simple and common switch Figure 1-21, is a disconnect switch on a furnace or house lighting circuit. This has one source and one load. Even though several lights are on the same circuit, the switch is supplying power to only one circuit. This is called a *single pole, single throw* (SPST) switch.

NOTE   In the switch illustrations and all later wiring diagrams, an arrow is shown on the H (and later N) wires to indicate the *direction* to the *source*. The arrow has *no* relationship to the *direction* of *current flow*.

The heating/cooling switch on a thermostat has one hot source but two load positions, heating or cooling, Figure 1-22. This is a *single pole, double throw* (SPDT) switch. It also may have an *off* position which is not counted as a load.

If there are two hot poles and only one load—as for an oven, where two 120v power sources are combined to run one 240v appliance—the switch would be a

FIG. 1-21.   Schematic of a single pole, single throw (SPST) switch.

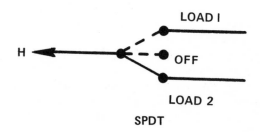

FIG. 1-22.   Schematic of a single pole, double throw (SPDT) switch.

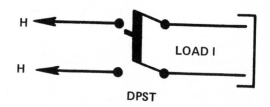

FIG. 1-23.   Schematic of a double pole, single throw (DPST) switch.

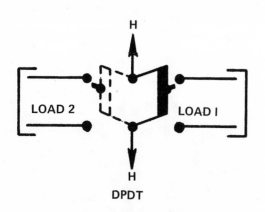

FIG. 1-24. Schematic of a double pole, double throw (DPDT) switch.

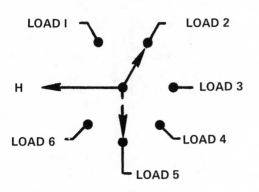

SP-MULTIPLE LOAD

FIG. 1-25. Schematic of a rotary switch shows that it is essentially a single pole switch with multiple throws.

double pole, single throw (DPST). See Figure 1-23.

If there are two power sources and two loads, the switch is *double pole, double throw* (DPDT). See Figure 1-24.

A switch may have one hot pole and many choices of loads, as in a rotary switch, Figure 1-25.

A slide switch, as in Figure 1-26, actually can be two SPST switches mounted together, with an *off* position. Note that in either position only one source and one load is connected.

Switches not only operate and control the circuit, but also serve as safety devices. The switch should always open on the line or power side of the fuse. If the fuse blows, power can be disconnected and the fuse replaced with no danger. Were the fuse on the line side, it would be impossible to change it safely.

The switch should *always* be in the hot leg ahead of the device it is supposed to control. This is an added safety factor when there is a short to ground.

With the switch ahead of the load, the load cannot operate with the switch open. When the switch is closed, and there is a direct short, the fuse will blow immediately, cutting off all power.

If a short to ground occurs between the load and the open switch, when the switch is in the neutral leg, the load will operate, the fuse will not blow and a serious shock hazard will exist. See Figure 1-27.

Switches of some kind are included in all electrical circuits of a heating and air conditioning system. They may operate on either line or low voltage.

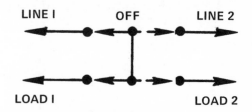

FIG. 1-26. Schematic of a slide switch shows that it is essentially two (SPST) switches mounted together with an off position.

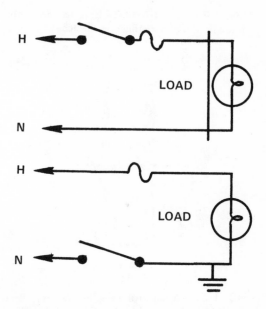

FIG. 1-27. With the open switch ahead of the load (above) the motor will not run. Since a direct short is present, the fuse will blow when the switch is closed. With the switch behind the load (below), a short to ground can cause the motor to operate with the switch open and create a serious shock hazard. Closing the switch will not blow the fuse.

TYPES OF SWITCHES

All switches can be *normally open*

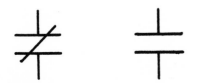

FIG. 1-28.   Electrical symbols for normally closed (left) and normally open switches.

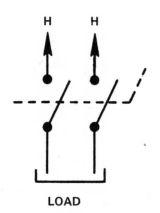

FIG. 1-29.   A manual switch is denoted by a dashed line through the schematic of the switch.

(NO) or *normally closed* (NC). The symbol for a NO switch is two parallel lines, Figure 1-28. A NC switch has the same symbol with the addition of a slanted line across the open space. This describes the position of the switch under normal operation or without being activated. A switch is designed to let current flow or to cut it off, and the *normal* position shows whether that particular circuit is normally energized or not.

In order for a switch to do its job, it must perform a mechanical action: that is, it must open or close a set of contacts. There are several ways in which this can be accomplished.

A manual switch, such as a furnace disconnect or the fan switch at the thermostat, is probably the simplest and most frequently used switch. As the name implies, the mechanical action, opening or closing the switch, is a manual operation and the switch will remain in the desired position until it is manually moved again. Any of the previously described actions (SPST, SPDT, etc.) can be accomplished with a manual switch. The switch symbol on a diagram usually will show a dashed line through the switch (or switches) to indicate that it is a manual switch, Figure 1-29. A manual switch will always, for safety reasons, break the circuit at the line contact end of the switch.

These switches, which may be either relays or solenoids, are operated by a magnetic coil constructed of wire wound around an iron plunger which is free to move up and down within the wire coil, Figure 1-30. When the coil of wire is energized, usually by low voltage, a magnetic

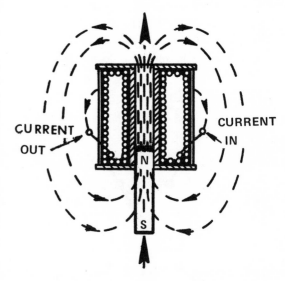

FIG. 1-30.   Movement of the plunger in an energized solenoid.

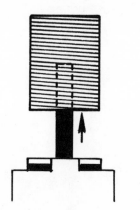

FIG. 1-31. Solenoid-operated NC switch.

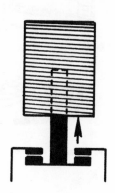

FIG. 1-32. Solenoid-operated NO switch.

field is set up which will pull the plunger up into the coil. Relays are equipped with electrical contacts at the end of the plunger which will *make* (close) or *break* (open) as the plunger moves upward, switching the second circuit which usually operates on line voltage.

In a normally closed (NC) switch, Figure 1-31, the plunger moves up, opening the electrical circuit, when the coil is energized. In a normally open (NO) switch, Figure 1-32, the upward movement of the plunger closes the circuit.

NOTE In this type of device, there are normally two *separate* and distinct electrical circuits, *independent* of each other. However, the coil circuit, when energized, controls whether or not the switch in the other circuit is open or closed, Figure 1-33. Relay and solenoid switches can be either normally open or normally closed and also can have any switching action: SPST, DPST, etc. Controlling line voltage with low voltage is a very common and desirable practice and the reasons for this will be discussed later under *Transformers*.

It should be noted here that a *switch*, in the true context, does not have to be a switch in an electrical circuit. When the coil activates the plunger, this could open or close a valve passage such as in a gas valve, Figure 1-34. The valve then shuts off or allows the flow of gas through the main valve.

Another method of actuating a switch is by *heat* or *temperature*. One type of heat-actuated switch uses a bimetalic

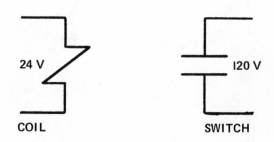

FIG. 1-33. The coil circuit, while independent of the switch, determines whether the switch is open or closed.

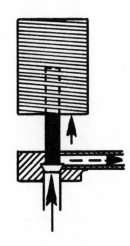

FIG. 1-34. Solenoid-operated gas valve.

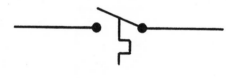

FIG. 1-35. A thermostat employs a heat-sensitive bimetallic switch that may be either normally open (above) or normally closed (below). *Courtesy General Controls ITT, Division of ITT Corp.*

element. (How this is accomplished will be discussed in detail under *Thermostats.*) At this point it can be noted that a bimetal changes position when the surrounding air temperature is heated or cooled. Electrical contacts placed on the end of a bimetal can be opened or closed by this action. A thermostat is a good example of a heat-actuated switch using a bimetal, Figure 1-35. A heat-actuated switch can be normally open (NO) or normally closed (NC) and can be opened or closed by either a rise or fall in temperature. They are generally single pole, single throw.

A fourth method of opening or closing a switch is by *pressure.* Pressure switches are commonly found on air conditioning systems because the compressor can be permanently damaged if system operating pressures exceed or go below the normal operating pressures. In a confined system, pressure is directly related to temperature, so the pressure controls are set to correspond to safe operating temperatures. A very common pressure control is a combination high-lo (dual) pressure control. The pressure in the system acts on a bellows which expands or contracts against a preset spring pressure. When the pressure overcomes the spring tension, it trips a mechanical linkage which opens a set of electrical contacts. Newer units have individual high and low pressure controls using the thermistor principle. These units sense temperature and trip when temperature reaches the corresponding pressure trip point. The sensing probes are soldered directly into the suction and discharge lines, Figure 1-36.

FIG. 1-36.   Hi-lo pressure control switch.

The fifth basic method of actuating a switch is by *light*, Figure 1-37. Certain materials, like cadmium sulfide, have considerable resistance to electrical current flow in the absence of light, but this resistance is reduced when light is present. Therefore, a grid of cadmium sulfide, called a Cad Cell, can act as a light-sensitive switch in the circuit. In the absense of light, no current will flow and the switch is, in effect, open. However, in the presence of bright light, such as an oil burner flame, the resistance diminishes and a small amount of current will flow, in effect, the switch is closed. How the Cad Cell is applied will be discussed in more detail later.

A final, specialized method of opening or closing a switch is by *moisture*, Figure 1-38. Certain substances, wood or human hair are examples, will swell with the addition of moisture and shrink as the moisture content is reduced. This principle is used in a humidistat which senses the moisture content in the air with its element and opens or closes a switch, operating a humidifier.

All of these various methods of opening and closing a switch can be used, alone or in combination, to control an electrical circuit and produce the desired results. It is important to remember, at this point, that *all* circuits will contain switches of some sort and that some circuits may have several switches controlling the same device.

TRANSFORMERS

Most of the control devices and many

FIG. 1-37. Cad Cell is a light-sensitive switch. *Courtesy Honeywell Inc.*

FIG. 1-38. Two types of moisture-actuated humidistats. *Courtesy Skuttle Mfg. Co.*

FIG. 1-39.   40 VA transformer. *Courtesy ITT Corp.*

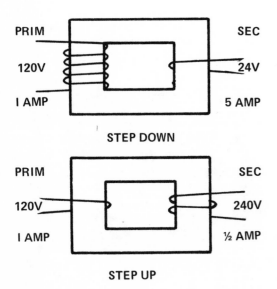

FIG. 1-40.   Outputs of step-down and step-up transformers.

of the safety devices used in residential heating and air conditioning operate on a low voltage circuit (24v) rather than line voltage.

There are a number of reasons for this:

- Low voltage wiring is much less expensive to run because wire size is smaller and easier to handle.

- Low voltage wiring does not require conduit and presents little, if any, fire or shock hazard.

- Components (switches, coils, etc.) designed for low voltage are much less expensive than similar components designed for line voltage.

In order to convert line voltage (120v) to low voltage (24v) a *transformer* is used, Figure 1-39. A transformer is merely an iron core with two sets of windings on opposite sides of the core. The line side of the transformer—that is, the side connected to the supply voltage—is called the *primary*. The output side is connected to the load and is called the *secondary*. A transformer which has 120v supplied to the primary and 24v at the secondary, is called a *step-down* transformer. That is, it reduces the voltage from 120v to 24v, Figure 1-40.

A transformer can also increase the voltage. For instance, a transformer with 120v on the line side which produces 240v on the load side is called a *step-up* transformer. That is, it increases the voltage from 120v to 240v. *Note that the primary is always the line side, but it is not necessarily the highest voltage.*

A transformer used to step-down the voltage from 120v to 24v will have 5 times the number of wire turns on the primary as on the secondary ($120 \div 24 = 5$), because the voltage is directly proportional to the number of wire turns. The current, or amperage, is inversely proportional. In the example of reducing voltage from 120v to 24v, the current (amps) will *increase* by 5 times. That is, 120v with 1 amp becomes 24v with 5 amps.

Transformers are rated by the power output of the secondary winding in terms of *volt-amps* (volts x amps or VA). This rating is very much affected by ambient heat, the applied load and the internal voltage drop. However, even though the load on the secondary changes, *the voltage will remain steady.* This is an important feature for a control or safety circuit.

Transformers must be sized correctly, that is, they must have enough power to handle the maximum load in the circuit. Like wire size, there will be a VA rating for each circuit and this is the *minimum* size transformer which can be applied. For example, if a 24v control circuit draws 1.4 amps of current, the transformer VA needed would be 24 x 1.4 or approximately 34 VA. A standard size 24v transformer is 40 VA, therefore this transformer could be used. As is true of sizing wire, a larger transformer can be used with no detrimental effect on the circuit. The secondary of many transformers is protected against overload by a fusible link which, if blown, requires the replacement of the entire transformer. Some have replaceable fuses and only the fuse needs be replaced. Primaries are protected by the fuses in the line.

METERS

Meters come in a variety of shapes and sizes, with some variations in scales. The two most commonly used in heating and air conditioning are a volt-ohmmeter and an ammeter, Figure 1-41. Other specialized meters will be presented later when their function is required for service procedures.

CAUTION: *All meters, while protected in many ways, are delicate instruments and can easily be harmed by misuse. Handle them carefully and follow directions.*

VOLT-OHMMETERS

All volt-ohmmeters will have two leads, one red and one black. The red lead is the positive (+) lead and there will be a jack on the front of the meter marked + or *positive*. The black lead is the minus (-) or *common* lead. The leads have an insulated shoulder which should be used to insert or remove the lead from the jack. *Do not pull the leads out by the wire— hold the insulated shoulder.* Some probes will have alligator clips on the end of both leads, others may have an alligator clip on the common lead and a straight metal probe on the plus (+) lead, while others may have a metal probe on both leads.

Most meters will have a main range switch for selecting the desired circuit: Resistance, AC Volts, DC Volts and sometimes Milliamperes. Each circuit, in turn, has as many as six ranges.

NOTE Some meters have the scales only and ac or dc is selected with a separate knob. Several voltage scales may be

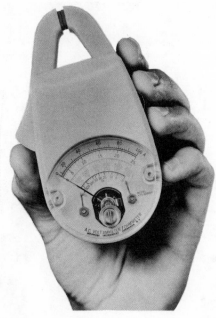

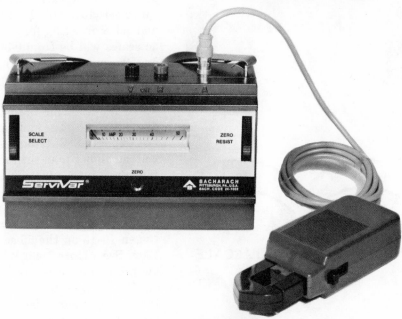

FIG. 1-41.   Ammeter (right). *Courtesy Amprobe Instrument.* Volt-ohmmeter (below). *Courtesy Bacharach Instrument Co., Division of AMBAC Industries.*

FIG. 1-42.   Face of volt-ohmmeter.

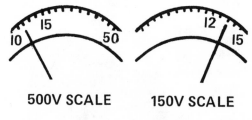

500V SCALE   **150V SCALE**

FIG. 1-43.   Indicating 120v on a 500v and 150v scale.

used, but the most common are multiples of 5 (1.5v, 5v, 50v, 150v, 500v and 1500v) or multiples of 3 (.6v, 3v, 60v, 120v, 300v, 600v and 3000v). With some meters, the probes must be moved to separate jacks to select the very low or very high voltages. Meters with scales in multiples of 3 are *not* recommended for heating and air conditioning work. Line voltages often run over 120v (122-126v). Using the 120v scale in this situation will burn out the meter. It is also difficult to get an accurate reading of 122v on the 300v scale.

There will be a minimum of 5 scales on the meter face: two for ac, two for dc and one for ohms, Figure 1-42. The ac scale is usually red and the others black. These scales will be calibrated to match the range selector knob. In the case of a scale and selector set up for multiples of 5, the ac scales will be from 0-50 (for ranges of 5v, 50v and 500v) and 0-15 (for ranges of 1.5v, 150v and 1500v). The range that is initially selected should be *at least* one range higher than the expected voltage, since high voltage can exist even where not expected. Putting a meter across a voltage higher than the top of the selected range will burn out the meter. The probes are placed on the H and N sides of the load. With the selector on the 500v range, a 120v source will move the needle to between 10-15 on the 50 scale, indicating 120v. See Figure 1-43. When the selector switch is moved to the next lower scale (150v), the needle will swing up to 12 on the 15 scale, again indicating 120v. Most accurate readings are obtained on the lowest scale encompassing the voltage being measured. Many meters will have a mirror behind the scale to eliminate parallax and give a more accurate reading.

Caution should be exercised when placing the probes in the equipment—particularly to avoid shorting (touching) the probe to any other metal. This can cause a direct short.

The dc scales are handled in exactly the same manner as the ac scales.

The Resistance (ohm) circuit usually has 3 ranges: Rx1, Rx100 and Rx10k. The x in this case means *multiplied by*, that is, the Rx1 scale reads resistance directly, the number read on the Rx100 scale must be multiplied by 100 to get the value, and the reading on the Rx10k scale is multiplied by 10,000 to obtain the value.

The resistance scale usually reads from 0 on the *right* to ∞ or infinity (no limit) on the left. When not activated, the needle will be at rest on the *left* hand side. Note that the meter will have a knob labeled *ohm adj.* This is to zero-in the scale for its most accurate reading. Place the selector in the Rx1 position and touch the two probes together. The needle should swing all the way to the right and stop at zero, Figure 1-44. If it stops short or goes beyond, turn the adjustment knob until the needle is exactly on zero. Ohms can now be accurately read for this range. This zeroing-in must be repeated for *each* range and should be checked first *everytime* a resistance reading is to be taken.

NOTE If the adjustment knob does not bring the needle to zero, this indicates

that the batteries in the meter are weak and should be replaced. The size and number of batteries will vary between meters, so the back should be removed and exact replacements selected.

AMMETERS

Ammeters are primarily used to check amperage. They may have lead attachments to read voltage and ohms, in the same manner as previously described for a volt-ohmmeter. One unique feature of some ammeters is that the scale is selected by a knob which positions the selected scale in a window in the handle, Figure 1-45. Only this scale is visible. Others have multiple scales similar to volt-ohmmeters.

An ammeter will have a set of jaws which can be opened by pressing a trigger

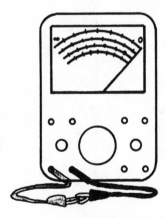

FIG. 1-44.   Zeroing-in an ohmmeter.

FIG. 1-45.   This jaw-type ammeter has a button that locks the needle on the reading. *Courtesy Amprobe Instrument.*

or lever on the handle. This allows the jaws to slip over and go around a wire in the circuit, without the necessity of disconnecting the wire. Amperage is read by encircling one wire of the circuit. Typical scales are 0-6 amp, 0-15 amp, 0-40 amp, 0-100 amp (and 0-300 amp on some). These scales are selected in advance. Again, always select the highest scale and move down to avoid damage to the meter.

Because amperage readings are sometimes taken in dark or inconvenient areas, many ammeters have a button on the handle which will lock the needle in its position when the reading is taken, and hold it after the probe is removed from the wire. On an Amprobe, this button is moved to the left to release and to the right to lock the needle in place. This feature allows measurements to be taken *in the blind.* The meter is then moved into the light where the scale may be easily read.

An amperage multiplier, that can be made by any serviceman, is an easy way to assure greater accuracy when measuring small amperages. Take some light wire and make a loop with exactly 10 turns. Place alligator clips on each end of the loop and tape the loop in 2 or 3 places to hold the wires together, Figure 1-46. This loop is attached in series to the circuit and the jaws of the ammeter are placed around the loop. The loop makes the value registered on the dial 10 times greater than the actual current. Using the smallest scale on the ammeter, and dividing the reading by 10, will give a far more accurate reading. The wire size or the diameter of the loop makes no difference.

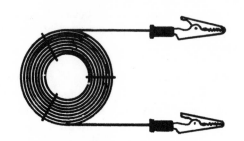

FIG. 1-46.   Amperage multiplier.

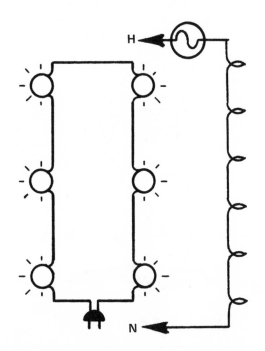

FIG. 1-47.   Pictorial and schematic diagrams of a string of Christmas tree lights.

SERIES CIRCUITS

In order for electricity to do any work, three things are necessary; a source, path for the current and a load (resistance). There are several ways for a load or several loads to be connected to the source. The first of these is the *series circuit.*

A series circuit is one in which the loads are connected end-to-end as in many of the older strings of Christmas tree lights. Each light or load is in a continuous sequence, beginning with the first and returning to neutral from the last. This is shown in both pictorial and schematic form in Figure 1-47.

When wired this way, the voltage starts at 120v. However, as it goes through each resistance, there will be a drop in voltage proportionate to the amount of the resistance. Upon return to neutral, the voltage will be zero. With a 120v source and 6 equal resistances, there will be a drop of approximately 20v across each resistance, Figure 1-48. If only 3 bulbs are used, they will each glow more brightly because each now has about twice the voltage drop (40v) to produce more light.

As long as there is a continuous circuit, it makes no difference which side of the resistance is wired to *H* and which continues to neutral, Figure 1-49. This fact should be kept in mind so that wires will be connected for the shortest runs and neatest appearance.

Current passes through each resistance in sequence, that is, from the first, to the second, to the third and so forth. There is no alternate path so, if one bulb is unscrewed, it acts like a switch and the

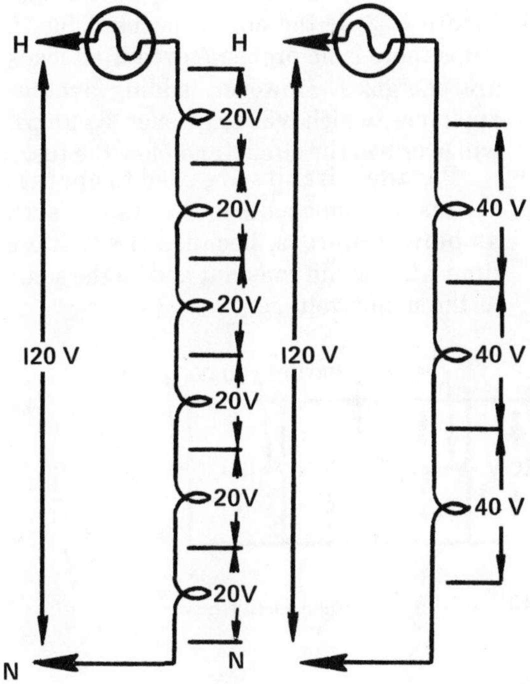

FIG. 1-48. 120v circuit with various resistances.

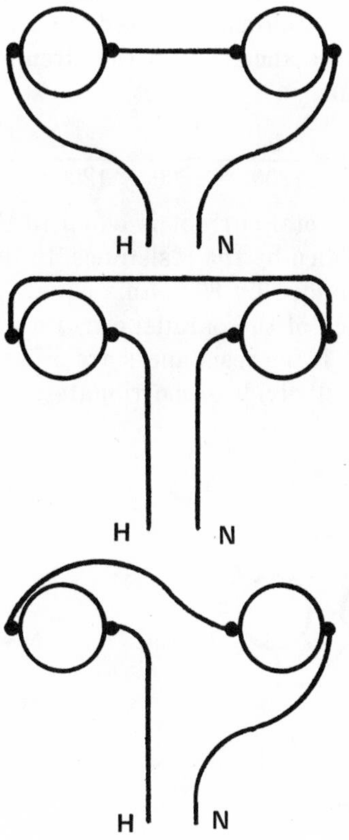

FIG. 1-49. Alternate methods of wiring in series.

entire circuit is de-energized. This is why several safety devices, controlling the same component, are wired in series with the component. If any one of them senses trouble, it will take the component off the line. *Switches are always wired in series and ahead of the device they control.*

If there are several unequal resistances in a series circuit, then the voltage drop will be greatest across the highest resistance. A 40-watt bulb will have smaller wire than a 100-watt, so the resistance of the 40-watt bulb will be much greater than the 100-watt bulb. If both are wired in series, most of the voltage will be used to light the smaller bulb and the larger one will not have enough voltage drop across it to light. This proves one of the basic principles of electron theory—*the larger the wire, the less the resistance.* It makes no difference which bulb is first in the wiring sequence. You can wire two light bulbs in series and prove this for yourself.

Also observe that, when the circuit is first energized, the bulbs will glow the brightest and will dim somewhat after a short period of time. This proves another theory—that a *cold wire has less resistance than a hot one.* As the bulbs warm up, they offer more resistance, causing less current flow so they do not glow as brightly.

The current in a series circuit is the same at all points in the circuit, regardless of the number and amount of the resistances.

PARALLEL CIRCUITS

A second basic way of connecting the loads to the source is by a parallel circuit. In this wiring method, the loads can be thought of as wires side by side (parallel) rather than end to end, Figure 1-50. In a parallel circuit, the voltage drop across each resistance, regardless of value, equals the full applied voltage. Each bulb will burn with full intensity and not dim as they did in a series circuit. This will remain true even if one circuit has a 100-watt bulb and another a 20-watt bulb, Figure 1-51. This is why all house lighting circuits are wired in parallel. Each circuit will have full voltage and the bulbs will light equally.

The total resistance of the parallel cir-cuit will always be *less than* the smallest individual resistance of the components. If the resistances in a parallel circuit are equal, the current (amps) will divide equally among the branches. For instance, if two bulbs, each with a resistance (R) of about 1200 ohms, are wired in parallel, the total resistance (R_t) in the circuit will be 600 ohms.

$$\frac{1}{R_t} = \frac{1}{1200} + \frac{1}{1200} = \frac{2}{1200} = \frac{1}{600}$$

The total current is equal to the voltage divided by the resistance. In this case, 120v divided by 600 ohms equals .2 amps and each of the parallel circuits will draw .1 amp. If the resistances are different, the amps will divide proportionately.

The total amperage in the circuit will be the sum of amps drawn by all the branch circuits. This is why the addition of another appliance to a circuit already near capacity will blow a fuse. Take a circuit with an air conditioner drawing 7½ amps and four other branches drawing 1 amp each, for a total load of 11½ amps, Figure 1-52. If the circuit is fused for 15 amps there is no problem, even if all loads are energized. However, adding another appliance, which will draw over 3½ amps, will overload the circuit and blow the fuse.

Parallel circuits are used to operate the major components in the system, such as blower motors, because the voltage drop will remain constant and be the same as the supply voltage.

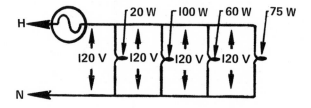

FIG. 1-51. All loads in a parallel circuit have equal voltage.

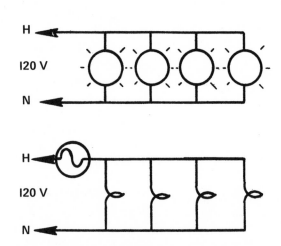

FIG. 1-50. Pictorial and schematic diagrams of a parallel circuit.

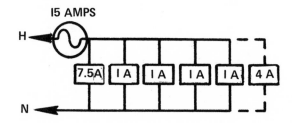

FIG. 1-52. Total amperage in a parallel circuit is the sum of the amps drawn by all the branch circuits.

COMBINATION CIRCUITS

In an actual installation, there will be combinations of series and parallel circuits. Some series circuits will be in parallel with each other and a series circuit may include components wired in parallel with each other. These circuits are called series-parallel or parallel-series.

A circuit is described as being in series or parallel, depending on its electrical relationship with another circuit or component. In Figure 1-53, circuit AC is a series circuit wired in parallel to series circuit BC. Both are in parallel to circuits DF and EG. A component may be wired in parallel with another component but still be part of a series circuit.

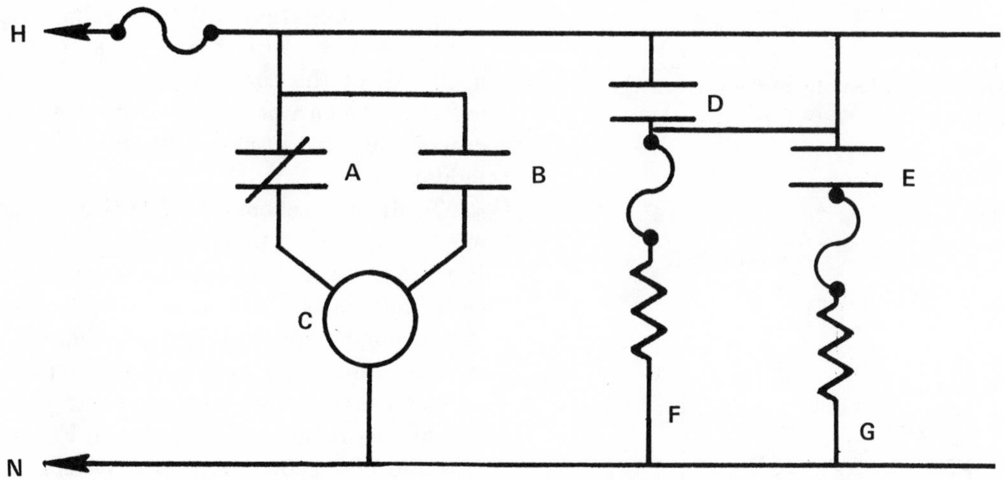

FIG. 1-53. Combination circuits have series circuits in parallel with each other.

CONTROL CIRCUITS

Control circuits are usually 24v circuits due to the ease of wiring, lower cost of components, safety from shock hazard and elimination of conduit. There are line voltage control circuits, but they are usually found on electric heat applications and larger commercial equipment. Low voltage circuits are sometimes drawn with a dotted line or lighter weight line to distinguish them from line voltage circuits.

Low voltage control circuits normally employ a relay coil to open or close a set of line voltage contacts. Relays were explained earlier. The low voltage coil will be in series with the actuating device; thermostat, limit, etc., and the line voltage contacts will be in series with the device to be controlled; blower, compressor, etc.

CIRCUIT DIAGRAMS

There are two basic methods of drawing a circuit diagram. One is to draw it with the components and wiring in about the same physical relationship as would be found on the actual installation. This is called a pictorial wiring diagram. The path of each wire is shown and the components are scaled to their approximate shape and location. This type of drawing is a good guide when actually wiring the equipment in the field and gives an accurate representation of the number and location of the wires required. Figure 1-54, for example, shows a common, heating-only circuit. A pictorial diagram is good for relatively simple wiring. However, as the wiring becomes more complicated, its value decreases due to the difficulty in routing and identifying individual wires. Further, from a service standpoint, it will not convey the interrelationship of components since it is difficult to tell which component is in which circuit.

A second way of drawing a circuit, called a schematic wiring diagram, is a solution to this problem. In this approach, the physical wiring is disregarded and the diagram becomes totally functional. Schematics always use symbols for the components, where pictorial diagrams usually show components by shape and relative size. A review of all the symbols learned thus far, plus some new ones, will aid you in reading schematic drawings.

A schematic diagram builds the draw-

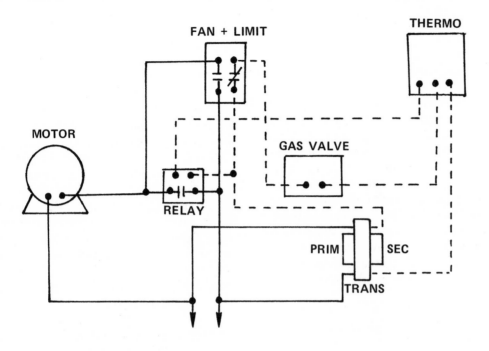

FIG. 1-54. Pictorial drawing of gas furnace controls.

FIG. 1-55. Schematic of the gas furnace control circuit shown in Figure 1-54.

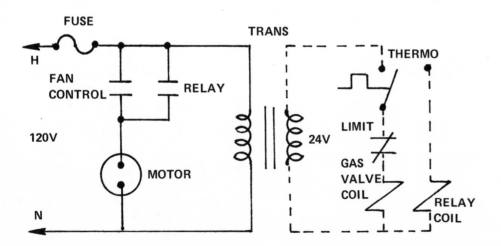

ing on a sideways *ladder*. That is, the line voltage circuit is represented as two horizontal, parallel lines. Components are shown connected between them, Figure 1-55. Line voltage is shown as a solid line and low voltage as a dotted line. (Note that some wiring diagrams use dotted lines to indicate field wiring rather than low voltage. Check the diagrams issued by the manufacturer.)

Note particularly that a relay may have its contacts in the line voltage circuit while the coil is shown, some distance away, in the low voltage circuit, even though these two elements are enclosed in the same housing. Separating the line and low voltage circuits makes the diagram much easier to *read*. Thus, the interrelationship of the components, and how each affects the other, is more easily determined. Examination will show what happens down the line if a given switch is open. All of the components being taken off the line can be easily identified because they are in the same *rung* of the ladder.

To draw a schematic, it is first necessary to identify which components are in the same circuit and whether or not succeeding circuits are in series or in parallel. A thorough understanding of schematics is most important in the more advanced service courses. The schematic and pictorial of the same circuit, shown in Figures 1-54 and 1-55, should be studied carefully.

240V POWER

120v, single phase power was shown to employ 2 wires: one *hot* (H) and one neutral (N). If two 120v, single phase power sources are linked with a common ground, then a 240v, single phase circuit is produced. This is a 3-wire service with the neutral (N) wire grounded at the source and two hot (H) wires run to the load. The third wire is grounded at the disconnect switch, Figure 1-56. A measurement across the two hot wires will read 240v. A measurement from each of the hot wires to the neutral will read 120v. This service is used to run larger motors and is alternating current just the same as the 120v source.

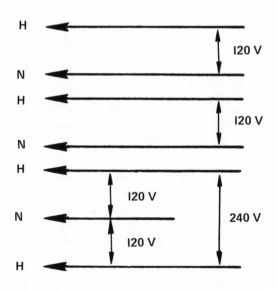

FIG. 1-56. Two 120v sources can be linked to provide 240v power.

Questions

1. What is electricity and where does it come from?
2. Describe the makeup of matter.
3. How do electrons move?
4. Give some examples of materials with a high resistance to electrical flow.
5. Name some materials that have a low resistance to electrical flow.
6. What other factors affect electrical flow?
7. What are the requirements for a complete electrical circuit?
8. What is the function of a fuse?
9. What is the difference between overcurrent and a short?
10. What happens when a motor is included in the electrical circuit?
11. Name five possible reasons for a fuse blowing.
12. What kind of a switch is the thermostat heating/cooling switch?
13. Give two reasons why switches are placed in the circuit.
14. Where is the switch located?
15. Name six ways by which a switch can be activated.
16. What is a step-down transformer?
17. How are transformers rated?
18. What is a volt-ohmmeter used for?
19. What color is the positive lead, the common lead?
20. Describe the scales available on a volt-ohmmeter.
21. What is the first thing to do when checking resistance?
22. What is an ammeter?
23. How is an ammeter used?
24. What is an amperage multiplier?
25. What is a series circuit?
26. Where is the series circuit used?
27. What is a parallel circuit?
28. When is a parallel circuit used?
29. Give four reasons for using low voltage circuits for control circuits.
30. What is the difference between a pictorial or physical wiring diagram and a schematic wiring diagram?
31. How many wires are there in 120v single phase power, and how are they marked?
32. How many wires are there in 240v single phase power, and how are they marked?

Review of Electrical Symbols

CAPACITORS		TRANSFORMER		SWITCHES CONT. PUSH BUTTON	CIRCUIT OPENING (BREAK)
COILS		THERMAL OVERLOAD		PUSH BUTTON	TWO CIRCUITS
CONTACTS	NO NC	THERMISTOR		MAKE BEFORE BREAK	
CONDUCTORS	CROSSING JUNCTION	CONNECTORS	MALE FEMALE	PRESSURE	
FUSE			4 CONDUCTOR ENGAGED	TEMPERATURE	CLOSE ON RISING
FUSIBLE LINK		SWITCHES DISCONNECT		TEMPERATURE	OPEN ON RISING
GROUND CONNECTION		SINGLE POLE SINGLE THROW	SPST	THERMAL RELAY	
LIGHT		SINGLE POLE DOUBLE THROW	SPDT	CONDUCTORS	
RESISTOR		3 POSITION	OFF SPDT	POWER (FACTORY) CONTROL (FACTORY) POWER (FIELD) CONTROL (FIELD)	
MULTIPLE CONDUCTOR CABLE		DOUBLE POLE DOUBLE THROW	DPDT	OR	
THERMOCOUPLE		PUSH BUTTON	CIRCUIT CLOSING (MAKE)	POWER (FACTORY) POWER (FIELD) CONTROL (FACTORY) CONTROL (FIELD)	

FIG. 1-57. Electrical Wiring Trainer. *Courtesy Allen Russell Associates.*

Problems

Having completed this chapter on Basic Electricity, the student should practice several different types of wiring, using simple, readily obtainable components. The problems that follow can all be demonstrated by mounting the components on Masonite or wallboard and wiring them as suggested. Components needed are 4 regular-size house light bulb sockets, an off-on switch with a fuse, 2 single pole, single throw switches, such as are used for light circuits, 2 single pole, double throw switches, a 120/24v, 20 VA transformer, a single pole, double throw blower relay, four 25-watt and two 7½ watt light bulbs, and sufficient 16 or 18-gauge wire to connect the various circuits. These components can be arranged on a board.

An alternate route, for those who would use these materials many times, would be Electrical Wiring Trainer, Model EW-1, available from Allen Russell Associates, P.O. Box 818, Marshalltown, Iowa 50158. This trainer, Figure 1-57, contains quick-connect banana plugs and supplies of wire so that individual wires do not have to be stripped and wired. Pads of the layout sheets are also available from the same source.

Either of these approaches will demonstrate the physical wiring of all the basic circuitry in a gas, oil or electric furnace, including thermostat and controls. The problems which follow give the step-by-step wiring procedure, since an understanding of the effects of the wiring are just as important as the actual wiring exercise. Once wired, the trainer will show, by energizing circuits and light bulbs, just what happens if a limit control goes out, thermostat calls for cooling, and so forth.

When talking about current flow in an electrical circuit, it is convenient to compare it to the flow of water. A load such as a light bulb has an upstream and downstream side, the same as a bridge spanning a river. The upstream side of the load is the side on which the current enters and the downstream side is the side on which the current leaves. Therefore, if we connect the upstream side of *Load A* to the source, and the downstream side of *Load A* to the upstream side of *Load B*, with the downstream side of *Load B* connected to neutral, we would have a series schematic circuit as shown in Figure 1-58. Note also that the *hot* and *neutral* wires have been indicated.

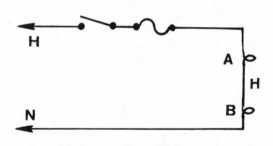

FIG. 1-58. *Load A* is downstream from the source, *Load B* is downstream from *Load A.*

When wiring the Electrical Trainer, the black wires should be used for the hot wires, white wires should be used for the neutral wires, and the blue wires should be used for the low voltage wires. The circuits should first be drawn on paper to assure that it is wired correctly before applying power, Figure 1-59.

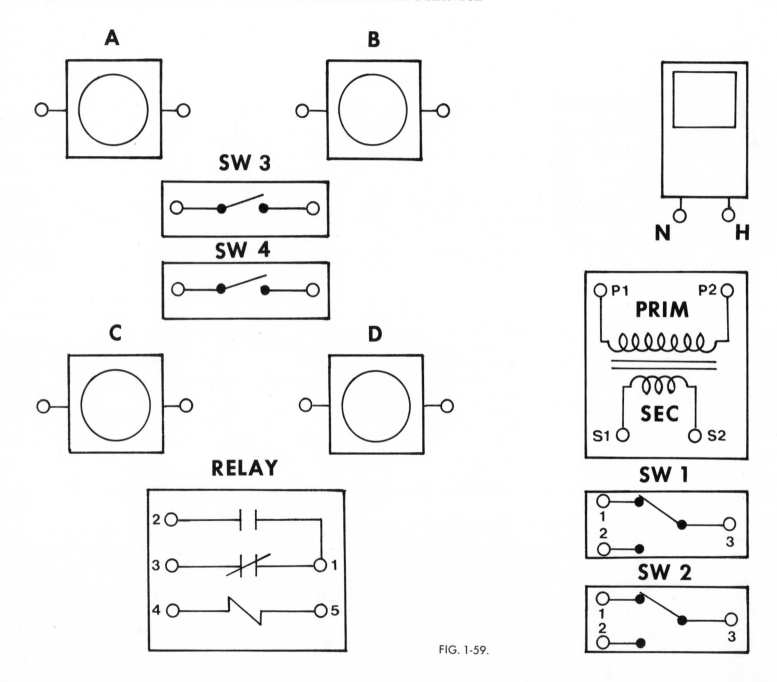

FIG. 1-59.

PROBLEM 1-1

Using Figure 1-59 as a guide, draw a pictorial wiring diagram of a series circuit.

A. Connect the hot leg from *H* to *Load A*.
B. Connect the downstream side of *Load A* to *Load B*.
C. Connect the downstream side of *Load B* to *Load D*.
D. Connect the downstream side of *Load D* to *Load C*.
E. Connect the downstream side of *Load C* to *Neutral N*.

When you have the pictorial wiring diagram drawn correctly, wire your circuit and answer the following questions:

1. Turn on the main disconnect switch and measure the following voltages and currents.
   VOLTAGES
   a. *H* to *N*.
   b. Across *Load A*.
   c. Across *Load B*.
   d. Across *Load C*.
   e. Across *Load D*.
   CURRENTS
   f. Between *Loads A* and *B*.
   g. Between *Loads B* and *D*.
   h. Between *Loads D* and *C*.
   i. Between *Load C* and *Neutral*.

   j. Were all of the voltages approximately the same? Explain.
   k. Were all of the currents the same? Explain.
2. Unscrew *Load A* and take the following voltage readings:

   a. Across *Load A*.
   b. Across *Load B*.
   c. Across *Load C*.
   d. Across *Load D*.

   e. Will current flow in any wire in this circuit? Explain.
3. Screw *Load A* back into the socket. Unscrew *Load C* and take the following voltage readings:
   a. Across *Load A*.
   b. Across *Load B*.
   c. Across *Load C*.
   d. Across *Load D*.

   e. Why is the full line voltage read across the unscrewed light bulbs in Questions 2 and 3?
4. Screw *Load C* back into the socket. Place a jumper wire across *Load D* and take the following voltage and current readings:
   VOLTAGES
   a. Across *Load A*.
   b. Across *Load B*.
   c. Across *Load C*.
   d. Across *Load D*.
   CURRENTS
   e. Between *Loads A* and *B*.
   f. Between *Loads B* and *D*.

   g. Why did the voltage drop get larger across *Loads A, B* and *C* and go to zero across *Load D*?
   h. Is the current the same throughout the circuit? Explain.
   i. Why did the current draw increase from the value in Question 1?
5. Draw a schematic of the circuitry in Problem 1.

PROBLEM 1-2

Draw a pictorial wiring diagram of the following series-parallel circuit.

A. Get a hot leg from *H* and connect it to *Load C*.
B. Connect *Load D* to the downstream side of *Load C*.
C. Connect the downstream side of *Load D* to *Neutral N*.
D. Connect *Switch 3* to the upstream side of *Load C*.
E. Connect *Load A* to the downstream side of *Switch 3*.
F. Connect the downstream side of *Load A* to *Neutral*.

When you have the pictorial wiring diagram drawn correctly, wire your circuit and answer the following questions.

1. With *Switch 3* in the open position, turn on the main disconnect switch and measure the following voltages and currents:
   VOLTAGES
   a. *H* to *N*.
   b. Across *Load C*.
   c. Across *Load D*.
   d. Across *Load A*.
   e. Across *Switch 3*.
   CURRENTS
   f. Between *H* and *Load C*.
   g. Between *Load C* and *D*.
   h. Between *Switch 3* and *Load A*.

   i. Why does *Switch 3* have the full line voltage across it?
   j. Is the current flow between *H* and *Load C* the same as between *Load C* and *D*? Explain.
2. Now close *Switch 3* and take these voltage and current readings:

VOLTAGES
a. Across *Load C.*
b. Across *Load D.*
c. Across *Load A.*
d. Across *Switch 3.*
CURRENTS
e. Between *H* and *Load C.*
f. Between *Loads C* and *D.*
g. Between *Switch 3* and *Load A.*

h. Why does *Load A* have the full line voltage across it?
i. Is the current flow between *H* and *Load C* the same as between *Load C* and *Load D*? Explain.

3. Unscrew *Load A* and take the following voltage and current readings:
VOLTAGES
a. Across *Load C.*
b. Across *Load D.*
c. Across *Load A.*
d. Across *Switch 3.*
CURRENTS
e. Between *H* and *Load C.*
f. Between *Loads C* and *D.*
g. Between *Switch 3* and *Load A.*

h. Why are all of the voltage readings identical across all of the loads in Questions 2 and 3, even though *Load A* is unscrewed in Question 3 and not in Question 2?
i. Why are the current readings in Question 3 different than in Question 2, even though the voltages are the same?

4. Draw a schematic of the circuitry in Problem 2.

PROBLEM 1-3
Draw a series-parallel pictorial wiring diagram as follows:

Line Voltage Connections
A. Connect the hot leg *H* to *P2* on the primary side of the step-down transformer.
B. Connect *terminal 1* of the relay to *P2* of the transformer.
C. Connect *terminal 2* of the relay to *Load D.*
D. Connect the downstream side of *Load D* to *P1* of the transformer.
E. Connect *P1* of the transformer to the *Neutral* leg *N.*
F. Connect *terminal 3* of the relay to *Load C.*
G. Connect the downstream side of *Load C* to the downstream side of *Load D.*

Low Voltage Connections
H. Connect *S2* secondary of the stepdown transformer to *terminal 3* of *Switch 1.*
I. Connect *terminal 2* of *Switch 1* to *terminal 4* of the relay.
J. Connect *terminal 5* on the relay to *terminal S1* on the transformer.
When you have the pictorial wiring diagram drawn correctly, wire your circuit and answer the following questions:
1. Place 100 watt bulbs in *Sockets C* and *D.*
Close *Switch 1* between *terminals 1* and *3*, turn on the main disconnect switch and measure the following voltages and currents:

VOLTAGES
a. *H* to *N.*
b. Across *Load C.*
c. Across *P1* and *P2* of the transformer.
d. Across *S1* and *S2* of the transformer.
e. Across *terminals 1* and *2* of the relay.
CURRENTS
f. Between *H* and *Transformer P2.*

2. Now close *Switch 1* between *terminals 2* and *3* and measure the following voltages and currents:
VOLTAGES
a. *H* to *N.*
b. Across *Load C.*
c. Across *Load D.*
d. Across *P1* and *P2* of the transformer.
e. Across *S1* and *S2* of the transformer.
f. Across *terminals 1* and *3* of *Switch 1.*
CURRENTS
g. Between *H* and *Transformer P2.*

h. Is the current reading the same or different between *H* and *Transformer P2* for Questions 1 and 2? Explain.
3. Draw a schematic of the circuitry in Problem 3.

Chapter 2
THERMOSTATS

The thermostat is possibly the most highly engineered precision component in the heating and air conditioning system. It is frequently taken very much for granted by the homeowner since it is set to the desired comfort level and then left to do its job. This section will primarily deal with low voltage (24v) thermostats, Figure 2-1. Line voltage (120v) thermostats are used in residential applications, mostly for electric heat, and all the basic principles discussed here will also apply to line voltage thermostats.

A thermostat has several important functions in the overall system operation:

1. The thermostat must control the furnace *on* and *off* times in order to maintain an even room temperature. It must have a wide range of settings to suit the individual comfort requirements of many types of people.

2. The thermostat must prevent over or underheating by assuring that the burners will not stay on too long nor stay off too long. By carefully and accurately monitoring the furnace *on* and *off* times, the thermostat can keep the spread between the coldest and warmest room temperature readings within 1° F. The range of room temperature is called temperature *swing*. For ideal comfort conditions, it is desirable to maintain a room temperature within the 1 or 2° range. The *swing* will depend upon the design conditions and the equipment selected. It can run to 6° F.

3. The thermostat must be able to maintain the desired room temperature over a very wide range of outside temperatures, since the heat loss from the living space will be accelerated by extremes of outside temperature. For instance, at 30° outside temperature, the furnace will have longer *off* times since the room heat loss will not be as great and the amount of heat put into the room will remain there for a longer period of time. By the same token, if the outside temperature is -20° F, the internal losses will be accelerated and the furnace might have to run all of the time to maintain even and uniform interior temperature conditions. Thus the thermostat has to sense and

FIG. 2-1.   Two typical low voltage (24v) thermostats. *Courtesy ITT-General Controls* (above), *Honeywell, Inc.* (below).

react to a great number of variables, including the inside room temperature, the outside ambient temperature and the amount of time the furnace actually runs.

4. The thermostat must automatically respond to all of these variables without any additional action by the homeowner, other than setting the comfort conditions desired within the living space.

5. The thermostat must overcome the effects of vibration and thermal inertia.

The main components of a thermostat are a thermometer, which senses the indoor room temperature, a dial for selecting the comfort level desired, plus a series of switches to select the desired fan control and to select either heating or cooling. It also has several internal components to compensate for other variables. How these components work together will be the subject of this chapter.

BIMETALS

It is a basic fact of physics that different metals, when heated or cooled, will expand or contract to a varying degree. Thus one metal, when heated, might expand considerably while a second metal, subjected to the same temperature and time of heating, will expand to a considerably less degree. This is the basic principle of a bimetallic device.

Brass, for instance, will react to heat and expand to a great degree, while a second metal, invar, will expand to a much lesser degree. If strips of these two metals are bonded together, it becomes a bimetal-

lic strip. When this strip is heated, the brass will expand more than in invar, and distort the strip to form an arc with the brass on the outside or larger radius of the arc and the invar on the inside or shorter radius, Figure 2-2. When these metals are cooled, the brass will shrink considerably more than the invar. Therefore it will become smaller and form the shorter radius, curving the arc in the opposite direction.

If, instead of leaving both ends free to expand or contract, one end of the bimetal is anchored, then the free end will move up or down, depending upon whether it is heated or cooled. By using a scale at the

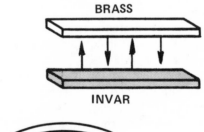

BRASS

INVAR

FIG. 2-2.   A bimetallic strip, formed by bonding brass to invar, will form an arc when heated or cooled due to unequal rates of expansion of the two metals.

free end, that is calibrated in degrees of temperature, it is possible to read the temperature of the surrounding air directly, Figure 2-3. Therefore the bimetal can become a thermometer, measuring the temperature surrounding it.

In addition to being a thermometer, a bimetal can be used as a switch. By replacing the calibrated scale with a set of

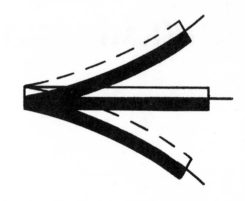

FIG. 2-3.   A bimetallic strip with one end anchored.

electrical contacts at the movable end of the bimetal, the bimetal can be made to *make* or *break* these contacts as it moves. See Figure 2-4. This allows the bimetal to open or close an electric circuit on either temperature rise or temperature fall, depending upon how the contacts are mounted. In other words, if it is desirable to close the switch as the temperature rises to a predetermined *set point*, the contact can be located at this point. When the temperature rises high enough, the bimetal will bend down and close the switch. In a like manner, if the temperature drops below the *set point*, the bimetal will bend up and

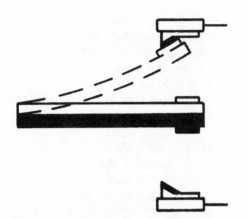

FIG. 2-4. A bimetallic strip can be used to *make* or *break* an electrical contact.

open the circuit. If it is necessary to *make* the circuit on temperature fall, then the location of the contacts is reversed. Thus a bimetal can start or stop another device in the circuit on either temperature rise or fall.

The direction and type of movement of the bimetal, as it is heated or cooled, is very much dependent upon the shape of the bimetal, Figure 2-5. Bimetals can come in many shapes, depending upon what mechanical action is required. For instance, it can be bent in a U-shape. With this configuration, the top ends will come closer together or spread farther apart, depending upon whether or not the bimetal is heated or cooled.

The bimetal can be wound in a spiral and anchored at one end. In this case, heating or cooling the bimetal will cause the spiral to lengthen or shorten. This shape is used in furnace limit controls.

A bimetal can also be made in the

shape of a cupped disc which will snap into either a convex or concave arc as it is heated or cooled. This shape is used in external compressor overloads.

Finally, the bimetal can be wound into a coil, like a watch spring. This shape will give a rotary motion at the unanchored end, as it is heated or cooled. This particular shape is the one normally used in a thermostat. The bimetal is a very thin, lightweight coil of metal which is very

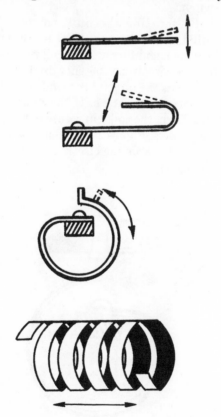

FIG. 2-5. Bimetallic elements are available in many shapes.

sensitive to changes in temperature. It will, therefore, respond very quickly to the temperature of the air around it.

The coiled bimetal can be anchored either in the middle or at the outer end. In thermostat applications, it is normally anchored in the middle with a pointer at the outer end, overlaid upon a scale calibrated in degrees, Figure 2-6. In this way, the coiled bimetal becomes a thermometer, with the pointer or end of the coil indicating the actual room temperature on the face of the thermostat.

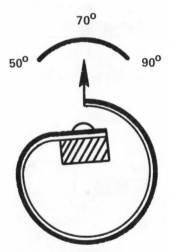

FIG. 2-6. A coiled bimetallic strip calibrated for a typical thermostat.

MERCURY SWITCHES

In addition to being used as a thermometer, the coiled bimetal can also be used to open and close another type of switch at a certain preset temperature. In addition to the temperature-indicating bimetal, a second bimetal coil within the thermostat is equipped with a mercury switch. This mercury switch is used to control the furnace *on* and *off* cycles.

A mercury switch is nothing more than a small glass enclosure with two wires or electrodes inserted in one end, Figure 2-7. The bulb also contains a bubble of mercury within it. Mercury is a liq-

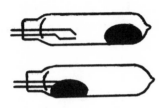

FIG. 2-7.   Mercury switch in the *off* and *on* positions.

uid metal and will flow back and forth in the bulb due to gravity. It is also a very good conductor of electricity. If the bulb is tipped away from the electrodes, the mercury will run to the opposite end and no contact will be made. The switch will then be open. If it is tipped toward the electrodes, the mercury will flow to that end and enclose both electrodes. This will allow electrical current to flow in the same manner as a manual or other type of switch.

By locating the bulb on top of the coiled bimetal, so that the mercury can flow back and forth as the coil rotates, the furnace circuit can be energized or de-energized as the bimetal responds to changes in room temperature. If the temperature goes down, the coil will rotate to the left as it contracts, Figure 2-8, tipping the mercury bulb towards the electrodes turning the furnace on.

As the temperature increases, due to the furnace supplying warm air to the living space, the coil will unwind in the opposite direction, tipping to the right. The mercury will flow away from the electrodes, breaking the circuit and stopping the furnace, Figure 2-9.

MAGNETIC SWITCHES

Another method of energizing and de-energizing the furnace circuit uses magnetic force to *make* and *break* contacts, instead of a mercury tube. In this system, the bimetal has an electrical contact on the end that moves with changes in temperature.

A second fixed contact, which includes a permanent magnet, is also located in the thermostat. As the moving contact approaches the fixed contact (room temperature falls below the comfort setting), the attraction of the magnet snaps the contacts together, completing the circuit and starting the furnace, Figure 2-10. To break contact, the bimetal must move in the opposite direction with a force sufficient to overcome the magnetic force holding the contacts together. For continued, dependable operation, this type of thermo-

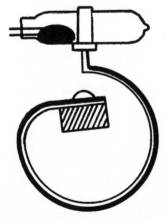

FIG. 2-8.   Mercury switch attached to a coiled bimetal (*on* position).

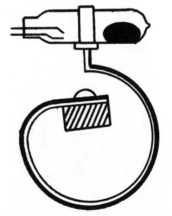

FIG. 2-9.   As temperature increases, the mercury switch is rotated to the *off* position.

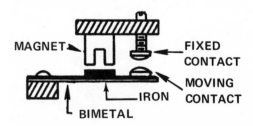

FIG. 2-10.   Magnetic switch.

stat must be kept free of dust or dirt inside the cover. Dirt can foul the contacts and weaken the magnetic field.

SETTING THE COMFORT LEVEL

The room temperature setting must be adjustable because different people like to have different comfort levels. Therefore a dial adjustment is provided in the thermostat to allow the homeowner to select the most satisfactory room temperature. If the dial is at 70° and the homeowner wishes to maintain 68°, he can merely rotate the dial adjustment to 68°, Figure 2-11. This dial adjustment is attached to the center or anchor point of the bimetal. Moving it to 68° rotates the center point slightly so that the bimetal is tilted a little further away from the heating or *make* point of the bulb. What this does, in effect, is to change the point (temperature) at which the thermostat will *make* and *break*. This new setting maintains the temperature selected by the homeowner. Burner cycles and running time will remain the same.

In a similar manner, if the homeowner wishes to maintain 72°, the dial adjustment can be turned up to the new temperature setting. This tips the coil in the opposite direction, changing the point at which the thermostat *makes* and *breaks*. Therefore the temperature in the room will be held at the desired higher level.

With this very simple adjustment, the homeowner can select the temperature which he wishes to maintain and hold it within very narrow limits. With continuous blower operation, the thermostat will normally hold the room temperature

within a range of plus or minus 1½°, with 8 to 10 cycles per hour. By employing short, but frequent, *on* times, the space temperature can be held within very narrow limits, even in very mild weather. The more frequent the *on* times, the easier it is to maintain this very small temperature *swing* above or below the *set point* of the thermostat.

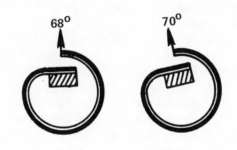

FIG. 2-11. The temperature level is adjusted by rotating the centerpoint of the bimetal.

DIFFERENTIAL

Under ideal conditions, a good thermostat will maintain the room temperature within a very narrow temperature range. However, because of the sensitivity of its very lightweight bimetallic coil, vibration, even that of walking across a room, might upset the thermostat enough to bounce the relatively heavy mercury bulb back and forth. This could cause the circuit to rapidly *make* and *break*, causing the furnace to cycle erratically and thus *chatter*. This action is hard on the furnace and the

electrical components, so it should be avoided if possible.

Cycling due to vibration may be overcome by building the mercury bulb with a slight hump in the middle, so that the normal gravity flow of the mercury is retarded, Figure 2-12. Before making or breaking the circuit, the mercury must run over this hump which slows it down.

FIG. 2-12. Thermostat bulb with hump for differential.

This creates a temperature differential that causes the furnace to turn off at a higher temperature point than where it turned on.

Another factor is the weight of the mercury itself. This additional weight on the very sensitive bimetallic coil requires the bimetal to twist further to *make* or *break* the circuit than would be required by the temperature alone. The combination of the weight of the mercury and the hump in the bulb produce a 2° temperature differential.

This means that, if the dial is set at 70°, the furnace will come on when the temperature falls to 70° and will keep running until the room temperature has reached 72°. This lag or differential will compensate for slight vibrations in the room and prevent the mercury in the bulb from *making* or *breaking* the circuit in an erratic fashion.

HEAT ANTICIPATION

When a heating device is turned on, it does not reach its maximum heat capacity immediately, but takes a certain amount of time to warm up. In a furnace, this is the time required to heat up the mass of metal in the heat exchanger and the duct-work, plus the air passing over the heat exchanger. All have to come up to the proper supply temperature before the room conditions can be maintained without drafts. This delay in delivering air at the proper temperature is known as *thermal inertia*, which can increase the temperature differential in the room even more.

Conversely, when the thermostat is satisfied and the burners turned off, all of these components (heat exchanger, ducts, etc.) retain a certain amount of heat which will be delivered to the room for a short period of time, even though the burners are off, causing the room to overheat.

Therefore, even though the thermostat signals the furnace to come *on*, there will be a time lag before the furnace will begin delivering air at the proper temperature. This allows the temperature of the room to fall below the desired comfort setting. Conversely, after the burners shut off, additional heated air will be delivered to the space causing the room to overheat beyond the point of the thermostat setting.

To overcome this thermal inertia, a heat anticipator is built into the thermostat which turns off the burners before the room actually reaches its temperature *set point*. By *fooling* the thermostat into believing that the correct room tempera-

ture has actually been reached, it turns off the furnace ahead of time and the residual or remaining heat in the heat exchanger and duct system will bring the conditioned space only slightly above the thermostat *set point*.

The heat anticipator is a small wire resistor, located close to the bimetal and wired in series with the gas valve or oil burner relay, Figure 2-13. Whenever these components are energized, the anticipator is also energized. Since it is a resistor, it will give off heat which will be added to the room ambient temperature sensed by the bimetal in the thermostat. This additional heat will cause the bimetallic coil to rotate more than it would in response to the actual room temperature and therefore turn off the burners before the room actually reaches the thermostat *set point*.

Experiments have determined that the human body can sense a change in temperature of about 1½° F. Thus, with a 2° differential, most people would sense this variation and be uncomfortable.

The heat anticipator compensates for this 2° temperature override built into the thermostat to overcome its sensitivity to vibration and offsets thermal inertia. This brings the thermostat back to the point where it can control room temperature to within less than 1°, since both the vibration problem and thermal inertia problem are overcome.

Some heat anticipators are a fixed resistance which is matched to the specific equipment to which that thermostat will be applied. In this case, no adjustments are possible. Most thermostats have an adjustable anticipator so that the resis-

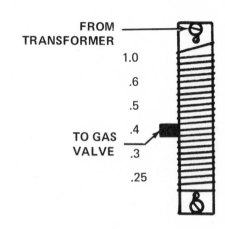

FIG. 2-13. The heat anticipator is essentially a resistor.

tance, and therefore the heat produced, can be varied with the application. The resistance is matched to the amount of current passing through the gas valve, oil burner relay, or other items in the low voltage control circuit so that the exact amount of heat will be delivered to the bimetal to compensate for the inherent differential. The amount of heat delivered by the heat anticipator will determine the length of the burner *on* cycle. The more heat that is supplied to the bimetal, the shorter the cycle. Less heat will produce a longer cycle.

The heat anticipator is merely a piece of coiled wire with a movable lever which selects the amount of resistance to be used. One end of the coiled wire (top in Figure 2-14) is connected to the transformer. The lever is connected electrically to the gas valve control circuit at the furnace. The total amount of heat produced by the anticipator depends upon the length of the wire selected (resistance) and the amount of current passing through it. The greater the length of wire selected (the lower the setting), the greater the resistance and the more heat produced by the anticipator. With the control circuit drawing .4 amps, if the selector lever is moved to .25 amps (more resistance), the anticipator will produce more heat and shorten the burner cycle.

There are several reasons for having an adjustable heat anticipator. One is that some people would like to have their burners cycle more or less frequently. Changing the amount of heat anticipation can accomplish this. Lowering the setting shortens the burner *on* cycle and raising it lengthens the cycle. Another reason is that different makes of gas valves draw different amperages. A higher amperage draw requires fewer turns of the resistor to generate the heat required to produce about a three-minute *on* cycle. In Figure 2-15, if the current draw is .6 amps, the lever would be set at this point. With a current draw of .35 amps, more turns are necessary. In both cases, the burner on cycle would be the same, since both settings will generate the same amount of heat.

There is another reason for an adjustable heat anticipator. Even though a gas valve drew its normal .20 amps, when additional control relays are added to the same circuit, they will change the amperage draw of the total circuit. Thus the heat anticipator must be able to adjust to this

and still generate enough heat to produce about a three-minute *on* cycle. Most gas valves are marked for .20 amperage rating and heat anticipator setting. This would

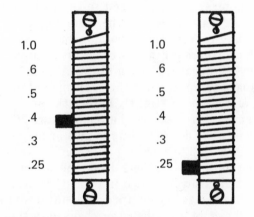

FIG. 2-14. Setting the anticipator at .40 amp produces less heat and longer *on* cycles, at .25 amp, the heat is increased and *on* cycles are shorter.

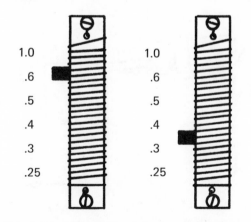

FIG. 2-15. Since both of these anticipators are matched to two differently rated gas valves, they will produce the same *on* cycle.

be the general set point in most up-flo furnace applications. The setting for a down-flo furnace might be as high as 0.45. How this value can be checked will be discussed in a later section of this text.

Most furnaces will be equipped with a sticker advising the installer or serviceman of the proper heat anticipation setting. Lacking this information or a means of checking the exact amperage draw, a general approximation would be to set a gas furnace at 0.25, an oil furnace at 0.45 and an electric furnace at from 0.20 to 0.40 (some however can run as high as 1.1).

Figure 2-16 shows a typical heating-only internal thermostat circuit with heat anticipation. The furnace blower will cycle on the furnace fan control with the burners.

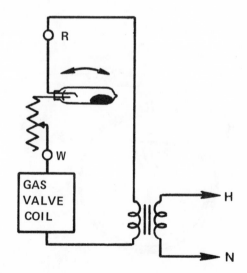

FIG. 2-16. Typical heating thermostat circuit with heat anticipator.

COOLING THERMOSTAT

In addition to a heating thermostat, which turns the furnace on when the temperature inside the house gets too low, it is also possible to have a cooling thermostat that turns on the air conditioner when the temperature gets too high. This can be easily accomplished by reversing the bimetallic coil so that, when it rotates due to an increase in heat, it makes contact and turns on the air conditioning system, Figure 2-17. As the temperature within the space comes down, the bimetal will contract and the mercury will flow to the opposite end, shutting off the air conditioner. Note that this bulb is built in exactly the same way as a heating bulb, with a hump in the middle which provides a 2° differential.

A cooling thermostat is also equipped with an anticipator since the principle of thermal inertia is the same for cooling as it is for heating. When the cooling system comes on, there is a time lag before the air temperature is reduced to the desired point. This is because the cooling system must first cool the evaporator, the air and ducts before it can deliver air at the proper temperature. To maintain an even space temperature, the cooling system must come on slightly before the space temperature reaches the temperature *set point*. The desired *swing* is plus or minus 2½° with 3 to 4 cycles per hour. The cooling anticipator differs from the heating anticipator, in that it is a fixed resistor (matched to the compressor contactor) and cannot be adjusted or changed in the field.

The cooling resistor is wired in parallel with its mercury bulb, Figure 2-18. When the mercury switch is closed, and the compressor is running, the resistance through the mercury switch is practically zero. Since current will take the path of least resistance, it will flow through the switch rather than through the anticipator. With no current flowing, the anticipator will not produce any heat. All of the current flows through the compressor contactor, pulling in the contactor and operating the compressor.

When the cooling demand is satisfied and the mercury switch opens, the resistor is then in series with the compressor contactor. This current is sufficient to produce heat in the resistor but, because of the large voltage drop across the resistor, there is not enough voltage remaining to pull in the compressor contactor, even though some current is flowing through it.

As with the heat anticipator, the cooling anticipator, on the *off cycle*, generates heat and *fools* the bimetal into thinking that the temperature in the space is higher than it actually is. This will tilt the mercury bulb, close the circuit and start the compressor before the temperature actually reaches the *set point*. Thus, the cooling anticipator compensates for the differential in the mercury switch and anticipates the need for cooling. This anticipation on both heating and cooling allows the thermostat to maintain the temperature in the space within very narrow limits, usually plus or minus 1° .

The effect of having current pass through the contactor, but not in sufficient quantity to pull it in, can be demonstrated. It may be recalled that when a 100 and 25-watt bulb were wired in series, the

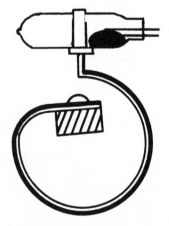

FIG. 2-17.   Cooling thermostat operation is the reverse of a heating thermostat.

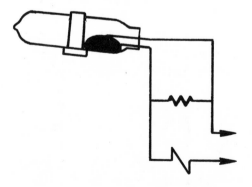

FIG. 2-18.   Cooling resistor wired in parallel with its mercury bulb.

100-watt bulb would not light because of the very small resistance, while the 25-watt bulb would light because it had a much greater resistance. The 100-watt bulb can be likened to the compressor contactor, and the 25-watt bulb likened to the cooling anticipator. Figure 2-19 shows a switch, representing the thermostat, wired in parallel with the anticipator (25-watt bulb). When this switch is energized (thermostat calling for cooling), then the 100-watt bulb, representing the compressor contactor, gets full voltage and will glow brightly, indicating that there is enough current to pull in the contactor.

Since the cooling anticipator starts the air conditioning system a little bit earlier than normal, it shortens the *off-time* of the air conditioning unit, which is important in the control of humidity. Humidity will be discussed later under Cooling, but most air conditioners are sized exactly or are slightly undersized to take care of this potential problem in the total comfort system.

HEATING-COOLING THERMOSTATS

Since one bulb can be used for controlling the heating system and the second bulb used to control the cooling system, thermostats can also be built with two bulbs, one for each system. If a system does use two bulbs, they can be mounted one on top of the other in order to combine them into a single thermostat for complete year-round control, as in Figure 2-20. This method is a little awkward and, of course, adds considerably to the weight for which the bimetal coil needs to compensate. Therefore, this is not a widely

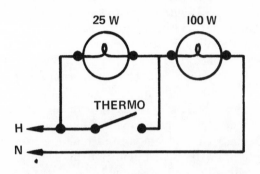

FIG. 2-19.   25W and 100W bulbs wired in series.

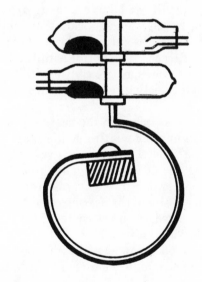

FIG. 2-20.   Two mercury bulbs combined to create a heating-cooling thermostat.

used method of building the heating and cooling thermostat.

A simpler way is to combine the heating and cooling functions into one mercury bulb. This can be done by putting the cooling electrode at one end of the bulb and the heating electrode at the other end, and adding a third or common lead, Figure 2-21. When the bulb is tipped in one direction, the mercury will complete the circuit through the cooling and common electrodes. When it is tipped in the opposite direction, it will complete the circuit through the heating and common electrodes. Note that, with this arrangement, it is not possible for the thermostat to call for heating and cooling at the same time.

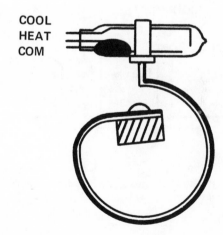

FIG. 2-21.   Typical combination heating-cooling thermostat.

The heating and cooling anticipators can be located immediately beside the bimetallic coil so that the coil will have the proper anticipator for the function selected. The basic wiring for each anticipator will be the same as that shown in the individual wiring diagrams, but both will be combined in one circuit.

With both heating and cooling in the same circuit, another potential problem must be solved. When the heating bulb is energized, for instance, the furnace will come on and the heating system will operate. When the heating load is satisfied, the bulb will switch back to shut off the heating system but, in doing so, will turn on 'the cooling system and reduce the heat in the space. The two bulbs will be constantly switching back and forth from heating to cooling, fighting each other. This is not at all desirable for proper temperature control and operation. The solution to this kind of a problem is to include a *heat-off-cool* switch which will put the system on either heating or cooling.

Figure 2-22 shows a typical heating-cooling internal thermostat circuit with both heating and cooling anticipation. This circuit will have a *heat-off-cool* switch to select either heating or cooling. For heating, the furnace blower will cycle on the furnace fan control with the burners. For cooling, the blower will cycle through a cooling relay with the compressor.

If the switch is in the heating position then, when the heating load is satisfied and the bimetal swings over to cooling, the cooling circuit will not be energized due to the open switch in the thermostat. In the cooling position, the heating circuit cannot be energized when the cooling load is satisfied. This switch therefore is a double pole, double throw switch with an *off* position in the middle where neither the heating nor cooling circuits can be energized.

SUBBASES

Many thermostats come in two parts: that is, the face and thermometer will be in one part and the second part, to which all the wiring is attached, will be the portion screwed onto the wall, Figure 2-23. This subbase, as it is called, will contain all the wiring connections to the thermostat and the screws or metal brackets attaching the face or cover will complete any electrical contacts required on the face. The subbase will contain the basic *cool-off-heat* switch as well as the blower switch. The use of a subbase allows the serviceman to get to all of the wiring without disturbing the basic thermostat. Note that many of the subbases will contain printed circuits to complete the internal wiring of the thermostat, and it is only necessary to connect the 4 or 5 thermostat leads to the designated terminals. Details for installation and thermostat wiring are always included with the thermostat that comes with the equipment and should be available to the serviceman.

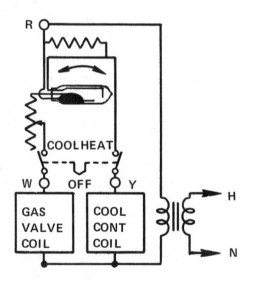

FIG. 2-22. Typical heating-cooling thermostat circuit with heating and cooling anticipation and an *off* switch.

Courtesy Honeywell, Inc.

FAN CIRCUIT

Many thermostats will also have a switch for the selection of indoor blower operation. Sometimes this is confusing since the selection may be marked *on* and *off; on* and *auto: cont* and *auto;* or *cont* and *int.* The *on* or *cont* position indicates that the blower will operate continuously, independent of the burner cycles. The *auto, off* or *int* positions mean that the blower will cycle with the burners.

In the *auto* or *off* position, most blowers will not come on until the burners have been on for several seconds. This allows the heat exchanger and ductwork to warm up before the blower delivers air to the room. The blower will remain running for some seconds after the burners are off, so that they can dissipate some of the heat in the heat exchanger, preventing it from getting too hot.

Without a blower control switch, the thermostat can only cycle with the burners on, through the fan control. To achieve continuous blower operation, additional wiring is necessary at the furnace.

Figure 2-24 shows a typical heating-cooling internal thermostat circuit with both heating and cooling anticipation. The circuit has a *heat-off-cool* switch plus a fan selection switch for *auto-on.* The switch is shown in the *auto* position, where the fan will cycle on both heating and cooling. In the *cont* position, it will run all the time.

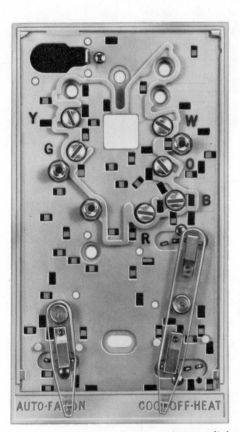

FIG. 2-23. Four typical thermostat subbases.

Most thermostat manufacturers label their terminals with the same letter coding which can correspond to the wire color used. Several widely used thermostats are listed below with their coding:

Manufacturer	Type	Common	Cooling	Heating	Fan
General	T91	V	C	H	F & G
General	T199	R	Y	W	G
Honeywell	T834	R	Y	W	G
Honeywell	T87	R	Y	W	G
Cam-Stat	T17	R	Y	W	G
Cont. Corp.	360	R	Y	W	G
White-Rodgers	IF56	RC & 4	Y	W	G

Some will have additional terminals for 2-stage heat (W_1 - W_2), two transformers (R_h and R) and other options. Figure 2-25 shows the internal wiring of a heating-cooling thermostat with fan control selection and the provision for using two transformers. Where only one transformer is used, the R and R_h terminals are *jumpered* and the thermostat will function like the one shown in Figure 2-24. Where two transformers are used, the jumper is removed. The internal switching (by selection of heat or cool) then uses one transformer for heating and the other for cooling.

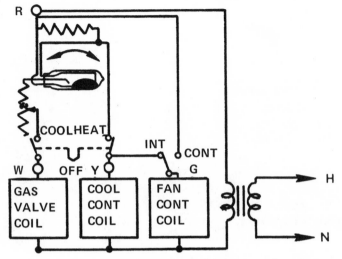

FIG. 2-24. Typical heating-cooling circuit with fan control.

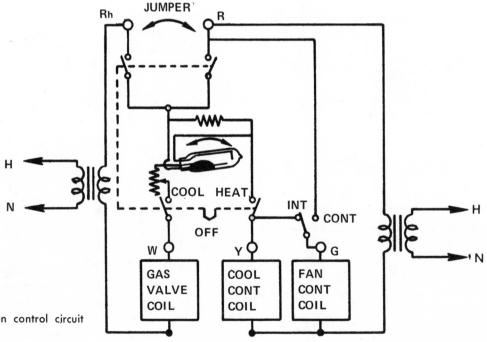

FIG. 2-25. Heating-cooling-fan control circuit using two transformers.

APPLICATION OF THERMOSTATS

The installation and application of a thermostat will have a considerable bearing on its accuracy and ability to maintain the comfort conditions desired by the homeowner. When installing a thermostat or checking out a complaint of poor comfort control, certain things must be kept in mind and checked:

• The thermostat must be leveled carefully, since the operation of the mercury switch depends upon gravity, Figure 2-26. If the thermostat is tilted one way or another, the accuracy of the dial adjustment setting will be upset and the thermostat will not function correctly.

• The thermostat should be located about 50 to 60 inches from the floor so that it senses the temperature at the height of a person's head when sitting. The air nearest the floor will be cooler and the air closest to the ceiling will be warmer, and this location gives a good average temperature for the overall comfort conditions.

• The thermostat should be located in the living space, usually in the living room, on an inside wall where it can sense the average temperature of the room.

There are many locations and applications of a thermostat which should be avoided, since all of them will result, in some way, in poor control of the average temperature within the room. Some mounting situations to be avoided are:

1. A thermostat should not be located on a wall which receives direct sunlight through a window or other opening since the radiant heat of the sun will raise the ambient temperature around the thermostat, causing it to misread the average room temperature.

2. A thermostat should not be located on the solid wall beside a doorway or appliance which could cause vibration. Constant vibration of the thermostat will throw it either out of calibration or out of control and, again, the comfort conditions in the room cannot be maintained.

3. A thermostat should not be screened by books or other objects which will interfere with the free flow of room air. If room air cannot get to the thermostat, it will not be able to sense the average temperature. Instead, it will only sense the temperature of the confined air around it and very likely give a false reading. In a similar manner, a lamp placed on a table close to the thermostat will heat the surrounding area and cause the thermostat to think the average room temperature is higher than what it actually is.

4. A thermostat should not be placed on a wall enclosing a duct run, chimney or other additional source of heat. The heat transfer through the wall will affect the sensing element of the thermostat and cause it to incorrectly read the average room temperature. The hole drilled in the wall, to wire the thermostat, should be sealed so that the thermostat is not affected by any drafts or other temperature variations within the wall space.

5. The thermostat should not be located on an outside wall. As with the wall in front of a chimney or duct run, the temperature of an outside wall will not be the room temperature, but rather will be affected by the heat loss or gain through the wall. This also will *fool* the thermostat into thinking the room temperature is higher or lower than it actually is.

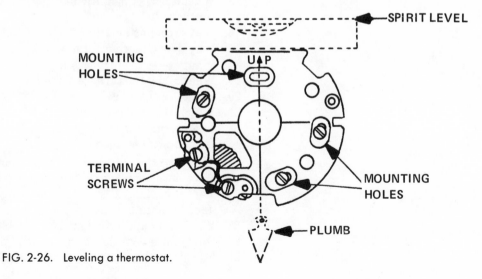

FIG. 2-26. Leveling a thermostat.

Questions

1. Name five important functions of the thermostat.
2. What are the main components of a thermostat?
3. What is a bimetal?
4. How is a bimetal used?
5. What direction will a bimetal move when wound into a coil and anchored at one end?
6. What is a mercury switch?
7. What is a magnetic switch?
8. What is meant by differential? Why is it important?
9. What is heat anticipation?
10. Why must a heat anticipator be built into a thermostat?
11. Is the heat anticipator in most thermostats fixed or adjustable? Why?
12. What is a cooling anticipator?
13. Is the cooling anticipator fixed or adjustable? Why?
14. When is the cooling anticipator energized? How?
15. When the cooling anticipator is energized, is there current through the compressor contactor?
16. Why is cooling anticipation important?
17. Why does a thermostat have a *heat-off-cool* switch?
18. What cycles the furnace blower on heating?
19. What cycles the furnace blower on cooling?
20. What is a subbase?
21. Why is the subbase used?
22. In a thermostat with the fan switch in the *auto* position, describe the blower operation on a call for heat.
23. Why does the *auto* position work this way?
24. Give the most usual designation for thermostat terminals:
    common_____, cooling_____, heating_____, fan_____.
25. What other options might be available?
26. Name three considerations for good thermostat application.
27. Name five thermostat mounting situations which are to be avoided.

Problems

PROBLEM 2-1

Using Figure 1-59 as a guide, draw a pictorial wiring diagram of the following series-parallel circuit. Use *Switch 2* as the heating-cooling portion of the thermostat, *Load A* as the gas valve and *Load B* as the cooling contactor coil. Connect the circuit as follows:

Line Voltage Connections

A. Connect the hot leg H to P2 of the transformer.

B. Connect *P1* of the transformer to *Neutral N*.

Low Voltage Connections

C. Connect *S2* of the transformer to *terminal 3* of *Switch 2*.

D. Connect *terminal 1* of *Switch 2* to *Load B*.

E. Connect the downstream side of *Load B* to *terminal S1* on the transformer.

F. Connect *terminal 2* of *Switch 2* to *Load A*.

G. Connect the downstream side of *Load A* to the downstream side of *Load B*.

After you have drawn the pictorial correctly, wire your circuit and answer the following questions:

1. Place 7½ watt bulbs in *Sockets A* and *B*.
   Place *Switch 2* in the off position, turn on the main disconnect switch and take the following voltage readings:
   VOLTAGES
   a. Across *H* and *N*.
   b. Across *S1* and *S2*.
   c. Across *1* and *3* of *Switch 2*.
   d. Across *2* and *3* of *Switch 2*.
   e. Across *1* and *2* of *Switch 2*.

   f. Why is there a voltage reading across *Switch 2* from *terminals 1* to *3* and *2* to *3*, but not from *1* to *2*?

2. Close *Switch 2* between *terminals 2* and *3* and take the following voltage readings:
   VOLTAGES
   a. Across *H* and *N*.
   b. Across *S1* and *S2*.
   c. Across *1* and *3* of *Switch 2*.
   d. Across *2* and *3* of *Switch 2*.
   e. Across *1* and *2* of *Switch 2*.
   f. Across *Load A*.
   g. Across *Load B*.

   h. Why is there a voltage reading across *Switch 2* from *terminals 1* and *3* and *1* and *2*, but not from *terminals 2* and *3*?
   i. Why is there a voltage reading across *terminals 1* and *2* of *Switch 2*, but not across *Load B*?

3. Close *Switch 2* between *terminals 1* and *3*, and take the following voltage readings:
   VOLTAGES
   a. Across *H* and *N*.
   b. *Across S1* and *S2*.
   c. Across *1* and *3* of Switch 2.
   d. Across *2* and *3* of *Switch 2*.
   e. Across *1* and *2* of *Switch 2*.
   f. Across *Load A*.
   g. Across *Load B*.

   h. Why are *Load A* and *B* very dim when either is energized?
   i. Is *Switch 2* in series or parallel with *Load A*?
   j. Are *Load A* and *B* in series or parallel?

4. Draw a schematic of the circuitry in Problem 2-1.

PROBLEM 2-2

Draw a pictorial wiring diagram of a typical gas furnace with cooling:

Switches 1 and *2* will be the thermostat, *Switch 3* will be a high temperature limit, *Switch 4* will be a fan control. *Load A* is the gas valve, *Load B* the cooling contactor coil and *Loads C* and *D* the two speed blower motor, with *C* low speed and *D* high speed.

Wire the circuit as follows:

Line Voltage Connections

A. Connect the hot leg from *H* to *P1* on the transformer.

B. Connect *P1* to *1* on the relay.

C. Connect *3* on the relay to *Switch 4*.

D. Connect the downstream side of *Switch 4* to *Load C*.

E. Connect the downstream side of *Load C* to the downstream side of *Load D*.

F. Connect the downstream side of *Load D* to *Neutral N*.

G. Connect *2* on the relay to the upstream side of *Load D*.

H. Connect *P2* on the transformer to *Neutral N*.

Low Voltage Connections

I. Connect *S2* of the transformer to *terminal 3* on *Switch 2*.

J. Connect *S2* of the transformer to *terminal 1* on *Switch 1*.

K. Connect *terminal 2* of *Switch 1* to *terminal 1* of *Switch 2*.

L. Connect *terminal 3* of *Switch 1* to *terminal 4* on the relay.

M. Connect *terminal 5* of the relay to *S1* on the transformer.

N. Connect *terminal 2* of *Switch 2* to *Switch 3*.

O. Connect the downstream side of *Switch 3* to *Load A*.

P. Connect the downstream side of *Load A* to *terminal S1* on the transformer.

Q. Connect *terminal 1* of *Switch 2* to *Load B*.

R. Connect the downstream side of *Load B* to the downstream side of *Load A*.

After you have drawn the pictorial correctly, wire your circuit and answer the following questions:

1. Place 7½-watt bulbs in *sockets A* and *B*, and 100-watt bulbs in *sockets C* and *D*.

    Close *Switch 1* between *terminals 2* and *3*.

    Close *Switch 3* and open *Switch 4*. Put *Switch 2* in the off position.

    Turn on the main disconnect and take the following voltage readings:

    VOLTAGES

    SWITCH 1

    a. Across *terminals 1* and *3*.

    b. Across *terminals 2* and *3*.

    c. Across *terminals 1* and *2*.

    SWITCH 2

    d. Across *terminals 1* and *3*.

    e. Across *terminals 2* and *3*.

    f. Across *terminals 1* and *2*.

    g. Across *Switch 3*.

    h. Across *Switch 4*.

    RELAY

    i. Across *contacts 1* and *3*.

    j. Across *contacts 1* and *2*.

    k. Across *coil 4* and *5*.

    l. Across *Load A*.

    m. Across *Load B*.

    n. Across *Load C*.

    o. Across *Load D*.

    TRANSFORMER

    p. Across *P1* and *P2*.

    q. Across *S1* and *S2*.

2. Close *Switch 1* between *terminals 1* and *3* and explain what happened.

3. Open *Switch 1* between *terminals 1* and *3* and close *Switch 2* between *terminals 2* and *3*, and explain.

4. Open *Switch 3*. Why did *Load A* go off?

5. Close *Switches 3* and *4*. Why did *Load C* come on when *Switch 4* was closed?

6. With *Switches 3* and *4* closed, close *Switch 1* between *terminals 1* and *3*. Explain what happened.

7. Close *Switch 1* between *terminals 2* and *3* and then close *Switch 2* between *terminals 1* and *3*. Why does *Load D* light and *Load C* go out?

8. Draw a schematic of the circuit in Problem 2-2.

Chapter **3**
HEATING
FUNDAMENTALS

Before considering the specific components of a heating unit, it will be necessary to understand some basic fundamentals, such as the nature of heat, upon which all furnace design is based.

Heat is a form of energy just like light, electricity and fuel. There is a rule called *The Law of Conservation of Energy* that states that energy in any form can neither be created or destroyed, but one form of energy can be transformed into other forms of energy, and energy can be transposed from one place to another. Fuel, as a form of energy, can be burned and transformed into two other forms of energy: heat and light.

There is no such thing as cold; cold is merely the absence of heat or, more accurately, a lesser heat content. There is some heat content in the air all the way down to absolute zero, about -460° F.

HEAT MEASUREMENT

Heat is always associated with matter. Matter is made up of very small particles called molecules which are constantly in motion. This motion is called *kinetic energy*. The speed of movement of the molecules determines the *state*, that is, solid, liquid or gas, of the substance. In a solid the molecules merely vibrate, in a liquid they move freely, and in a gas they move rapidly. See Figure 3-1. When heat energy is added to a substance, it increases the motion of the molecules and thus their kinetic energy. This motion (measure of kinetic energy) can be sensed by touch, and the measurement is called *temperature*. Temperature, therefore, is a measure of the *intensity* of heat.

The basic reference points for temperature measurement are the points at which water (matter) changes from a solid to a liquid and again from a liquid to a solid. These reference points are the same all over the world, if taken at sea level, and so can be duplicated anywhere. Sea level is chosen because atmospheric pressure is a constant (14.7 psi). Higher altitude (less pressure) will have a major effect on the point at which water boils, but this will be disregarded here.

On the Fahrenheit Scale (symbol, ° F) water freezes at 32° F and boils at 212° F. These then become the fixed reference points and there are 180 equal divisions or degrees between them to measure the temperature change, Figure 3-2.

On the Centigrade Scale (symbol, ° C) the reference points are 0° C for freezing and 100° C for boiling, with 100 equal di-

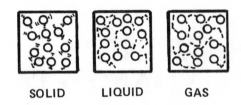

SOLID   LIQUID   GAS

FIG. 3-1. Molecular motion in a solid, liquid, gas.

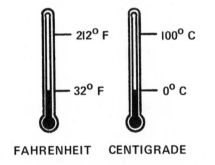

FAHRENHEIT   CENTIGRADE

FIG. 3-2. Fahrenheit and centigrade thermometers.

visions or degrees between them.

Heat energy is the *quantity* of heat contained in a substance and is measured in British thermal units or Btu's. A Btu is defined as the amount of heat required to raise the temperature of 1 pound of water 1° F. Btu therefore measures the total heat content of a substance by volume. A simple example: If there is a container with 1 pound of water at 70° F, and another container with 10 pounds of water at 70° F, they both have the same temperature but the 10 pound container would have 10 times the heat energy. Raising the temperature of 1 pound of water to 80° F would require 10 Btu's (10° x 1 lb x 1 Btu/lb), but it would take 100 Btu's to raise the temperature of the 10 pounds of water from 70° to 80° F (10° x 10 lbs x 1 Btu/lb). One Btu is about the amount of heat produced by burning a wooden kitchen match.

THE COMBUSTION PROCESS

Combustion is the process of burning a fuel (oil or gas in the case of a furnace) to produce two other forms of energy: heat and light. By observation and measurement, it is possible to determine the efficiency of the combustion process.

The combustion process is often taken for granted when, in fact, it is the key to good furnace design and eventual customer satisfaction. Simple adjustments to furnace combustion can frequently produce much better comfort levels and more economical operation. All service calls should include an inspection and check of the flames in actual operation. The combustion process must be thoroughly understood to perform this service.

The so-called *combustion triangle*, Figure 3-3, illustrates the three things that *must* be present in order to have a flame. Note that *all three* must be present.

1. *Fuel*   Any substance that will burn can be used, but in residential heating, natural gas, LP gas or oil will be the fuels.
2. *Heat*   Each substance has a different *ignition point*, that is, the temperature at which it will ignite and burn. Therefore, there must be enough heat supplied to raise the fuel's temperature to this point. A pilot light (constantly burning) or spark is usually used to produce the ignition heat.
3. *Oxygen*   Oxygen must be present in sufficient quantity to support combustion and, since it is consumed in the process, must continually be replaced. It can be supplied by forced air (air drawn in by a blower) or naturally from the surrounding air.

The need to have oxygen present for combustion is readily demonstrated by trying to burn fuel oil in an open pan. Oil is a combustible and a match or other flame can easily bring it to the ignition point. However, oil is heavy and, without being atomized (mixed with air), it will not burn. Even if a very high flame causes it to ignite, it will immediately burn itself out due to lack of oxygen.

Once all these elements are present in the required degree and amount, a flame can be produced and maintained.

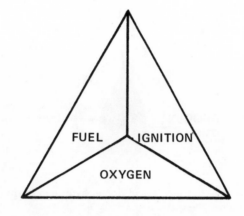

FIG. 3-3.   The combustion triangle. Fuel, oxygen and ignition are all required for combustion.

FUELS

There are a number of common fuels used in residential heating. To adjust their burners for proper performance, the serviceman must understand each type.

The most widely used fuel is natural gas, which is about 99% methane. Methane, composed of carbon and hydrogen, is one of many hydrocarbons which are excellent fuels. Its heating value ranges from 950 to 1125 Btu per cubic foot. Under most circumstances, it is considered to be about 1000 Btu/cubic foot. A furnace rated at 110,000 Btuh (Btu per hour) will therefore burn 110 cubic feet of gas per hour. Natural gas is a relatively slow burning gas with a maximum theoretical flame temperature of 3550° F. It has a specific gravity of 0.65 which means it is lighter than air, whose specific gravity is 1.00.

Another gas found occasionally is *manufactured* or *towns gas*. This is a by-product of other substances and is composed of hydrogen, methane, carbon monoxide and small quantities of carbon dioxide, oxygen and nitrogen. Its heating value is 500 to 550 Btu per cubic foot, about half that of natural gas. This is a relatively fast burning gas and the speed of burning increases as the hydrogen content goes up. Some manufactured gases are referred to as *snappy* gases, because of high hydrogen content, and special burners are required to handle them quietly. The maximum theoretical flame temperature is about 3560° F and its specific gravity averages about 0.45.

Some areas pipe in natural gas and mix it with manufactured gas produced locally. The mixed gases will average from 700 to 900 Btu per cubic foot. They will have a maximum flame temperature and specific gravity very close to natural gas. Usually a different orifice size is the only change necessary to have satisfactory combustion.

Another common gas, particularly in rural areas or sites not served by natural gas pipelines (such as mobile homes), is liquified petroleum gas or LPG. Two common ones are butane and propane. These are straight hydrocarbons compressed into a liquid state. They create their own vapor or gas pressure when their temperature is above the boiling point (butane +31° F, propane -43° F), and produce about 21,600 Btu per pound in the vapor state. Vaporized butane has a specific gravity of 2.01 (twice as heavy as air) and contains 3200 Btu per cubic foot. Propane has a specific gravity of 1.52 and contains 2500 Btu per cubic foot when vaporized. Flame temperature of both is about 3650° F.

Sometimes LPG gases are mixed with manufactured gases or air. Fortifying them this way can result in a specific gravity of less than air, although they may be diluted to the point where they contain only 550 to 550 Btu per cubic foot. This is important because any gas heavier than air requires special safety controls including 100% shutoff. This will be explained in more detail later.

Oil is distributed as No. 1 or No. 2 or kerosene. The most commonly used is No. 2 fuel oil (diesel oil) which has a heat content of 144,000 Btu per gallon. An oil furnace rated at 144,000 Btuh input will therefore require a pump capable of delivering a minimum of 1 gallon per hour.

PRODUCTS OF COMBUSTION

The basic elements present in all fuels are hydrogen and carbon atoms in varying amounts. They are called hydrocarbons. Hydrogen mixes quite readily with air and burns at a faster rate and at a lower temperature than carbon. It has a bluish color and burns first, using the air it needs for complete combustion, Figure 3-4.

The carbon particles are not completely burned until they reach the outside edge of the flame where they can obtain the required amount of air for complete

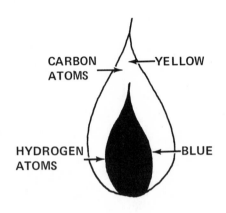

FIG. 3-4.   In the typical flame, hydrogen burns first, then the carbon.

combustion. The unburned carbon particles, slower burning but at a higher temperature, produce the bright light normally associated with an open fire.

If an open fire is examined closely, there will be a small blue flame at the base (caused by the hydrogen) and a much larger area of yellow which is the unburned carbon. Since the hydrogen burns first in the inner core, it uses the available oxygen and there is no additional oxygen present within the flame to completely burn the carbon. Thus the carbon does not burn completely until it reaches the outer edge of the flame where it can obtain the required oxygen.

During combustion, a chemical change takes place and most of the elements are rearranged. Note that the fuel is not destroyed since it is a source of energy and can be transformed but not destroyed. The fuel has hydrogen (H) and carbon (C) atoms while the air contains oxygen (O) and nitrogen (N) atoms. The products of *complete* combustion therefore become carbon dioxide (carbon + oxygen or CO_2) and water (hydrogen + oxygen or H_2O). The nitrogen from the air passes through the combustion process unchanged. None of these products are harmful. While different amounts of air are required for complete combustion of gas or fuel oil, the resulting products of combustion are the same.

$$H + C + O + N = CO_2 + H_2O + N$$

Complete combustion of any fuel requires approximately 10 cubic feet of air for every 1,000 Btu of heat contained within the fuel. In the case of natural gas, there are approximately 1,000 Btu's per cubic foot so, theoretically, 10 cubic feet of air is required for complete combustion of 1 cubic foot of gas. In actual practice, from 40 to 50% of excess air is desirable in a gas furnace. Therefore, good design would allow for between 14 to 15 cubic feet of air per cubic foot of natural gas. For theoretically complete combustion, 1 cubic foot of natural gas is combined with 10 cubic feet of air, plus the flame to achieve ignition, and the products of combustion become 8 cubic feet of nitrogen (air is approximately 80% nitrogen), 1 cubic foot of carbon dioxide and 2 cubic feet of water vapor.

Burning LPG, butane for example, which contains 3260 Btu per cubic foot, would theoretically require about 33 cubic feet of air. Actually, about 48 cubic feet of air are used (almost 5 times the air required for natural gas) to burn 1 cubic foot of gas. As will be shown later, this is the reason the burners for LPG must be designed differently than for natural gas.

For No. 2 fuel oil, the ratio becomes 1 lb of fuel oil plus 14.4 lbs of air (11.1 lbs nitrogen and 3.3 lbs oxygen) with a flame for ignition. Complete combustion yields 11.1 lbs of nitrogen, 3.2 lbs of carbon dioxide and 1.1 lbs of water. In terms of gallons, it requires 1,540 cubic feet of air for 1 gallon of fuel oil.

INCOMPLETE COMBUSTION

Incomplete combustion occurs when a gas flame cannot get enough oxygen to complete the combustion process. When the fuel is not completely burned, the products of combustion change from harmless elements to messy, irritable and possibly dangerous elements.

Incomplete combustion can also occur if a flame is cooled below its ignition point. This can be demonstrated by holding a cool glass ashtray over a candle, Figure 3-5. Immediately, soot will form on the bottom of the ashtray, indicating that part of the carbon did not burn. Soot is composed of semisolid particles of unburned carbon and is always an indication of incomplete combustion. As stated, it can occur by cooling the flame below its ignition point (called impingement of the flame) or from lack of sufficient air.

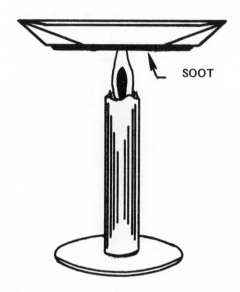
SOOT

FIG. 3-5. Cooling a flame below its ignition point creates soot.

The products of incomplete combustion from a yellow flame are as follows:
1. Carbon dioxide (CO_2)
2. Water vapor (H_2O)
3. Carbon monoxide (CO) - a deadly gas
4. Soot
5. Aldehydes (severe irritants to eyes, nose and throat)

FLAMES

A yellow flame is associated with an open fire; candle, bonfire, etc. The yellow color is produced by unburned carbon particles. These particles do burn completely as they reach the outer edge of the flame where they can obtain the additional oxygen required for complete combustion. A yellow flame requires a lot of space in which to burn and is subject to impingement if confined. Confinement results in incomplete combustion. However, an oil burner does burn with a yellow flame so its characteristics should be understood.

The oil is atomized in the nozzle, that is, mixed with air, which allows each drop to ignite. However, each droplet does not attain complete combustion until it reaches the outer edge and so oil burns with a yellow flame.

The occurrence of a blue flame within a yellow flame, due to hydrogen burning, has already been mentioned. However, there is another type of blue flame which has significant importance in gas heating. This blue flame results from mixing about half of the combustion air required for the burning process with the gas *before* it is ignited. The flame produced by this method is blue, rather than yellow, much smaller in size, has a higher temperature, gives out only a small amount of light and is easily adjustable. The total amount of heat produced by a yellow and a blue flame is about the same after burning equal quantities of gas.

The simplest demonstration of blue flame characteristics can be shown with a bunsen burner found in most science labs, Figure 3-6. In a bunsen burner, the gas is introduced under pressure at the bottom of an upright tube. Just above the base, there is a collar which has slots and a cover which can be rotated to open part or all of the slots. Opening the slots allows a quantity of air to mix with the gas on its way up to the top of the tube where it is ignited. This air is called *primary* air. Most burners in a gas furnace use this basic principle, even today, because it is efficient and easy to build.

With the proper amount of primary air, the flame has a bright blue inner cone surrounded by an outer mantle of light blue, Figure 3-6. This is the basic flame used in all gas furnaces.

Since combustion in this type of flame simultaneously occurs in two zones, on the surface of the inner cone and on the surface of the mantle, all of the carbon particles remain in a gaseous state until they reach the point of ignition. The inner cone is composed of a gas-air mixture rather than unburned carbon. Therefore, there is no yellow color indicating unburned carbon particles. The primary air, mixed with the gas, supplies enough oxygen to support combustion on the inner cone and the secondary air supplies the oxygen for complete combustion on the outer mantle.

As long as there is sufficient primary air, the flame will be blue with no unburned carbon. If the primary air is restricted, a yellow tip will appear on the cone due to the carbon particles not being completely burned until they reach the outer edge of the flame. If primary air is completely shut off, a yellow flame will appear, indicating that there is insufficient primary air.

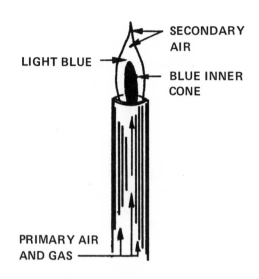

FIG. 3-6.   A bunsen burner uses primary air to maintain a blue flame.

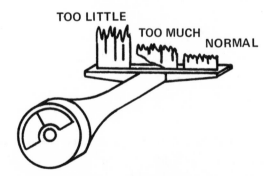

FIG. 3-7.   Effects of poor adjustment of primary air.

There are situations where a blue flame will have incomplete combustion and change to a yellow flame producing the harmful products previously listed. Causes of this are (see Figure 3-7):

Lack of Secondary Air   Too little secondary air, even though the flame is adjusted to receive a proper amount of primary air, will result in incomplete combustion. In this case, the outer shroud does not have enough oxygen to burn all the gas.

Excess of Primary Air   Too much primary air will distort the flame and lift it off the burner port. It will make a blowing noise and could blow itself out, just as a person blows out a match.

Inner Cone Impingement   If the inner cone impinges on a surface cooler than that required for ignition, then incomplete combustion will result as explained previously.

NOTE   Impingement by the outer mantle of a blue flame will not result in incomplete combustion. The heat for combustion is maintained by the inner cone and the outer mantle can still get enough secondary air to support complete combustion.

HEAT TRANSFER

Heat energy has one other characteristic which is used extensively in the heating and air conditioning industry. Heat will always flow from a warmer body to a cooler body. In other words, nature is constantly trying to equalize temperatures but never really succeeds. When the outside air is cold, man's body loses heat to the air so he must either put on more and warmer clothing, to restrict the flow of heat away from him, or he must raise the temperature of the air around him. Actually, he will do both. Raising the temperature around him is one of the prime functions of a furnace.

The reverse of this is also true. If the air around him is hot, he will absorb heat from the air. Now he must have additional body cooling (which he gets through perspiration) or he must lower the temperature of the surrounding air. This is one function of an air conditioning system.

Because heat energy will travel from the warmer body to a cooler body, this factor can be used to transport heat from one place to another. Three methods are used to accomplish this. They are:

1. Conduction   As with electricity, some materials will transport heat better than others. If one end of an iron rod is placed in a fire, the other end will become hot rather quickly. The heat energy travels through the rod from the hotter end (the one in the fire) to the cooler end. This is called *conduction*. Some materials, like wool and fiberglass, resist the passage of heat, which is why these materials are used for clothing or insulation. However, in all cases, heat will pass through them but at slower rates of travel.

This can be demonstrated by putting a quantity of boiling water in a cup. The cup will get warm and, if a metal spoon is placed in the cup, the opposite end will get hot, Figure 3-8. A wooden spoon, placed in

FIG. 3-8.   Conduction through a metal spoon.

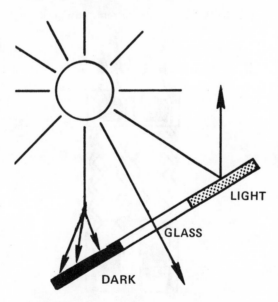

FIG. 3-9.   Radiant heat moves freely in space.

the same cup, will not get hot on the far end because wood resists the conduction of heat.

Conduction will also occur through different materials, including air, in contact with each other. The wall of a house can be composed of an air film on the inside wall, plasterboard or wood paneling, insulation, an interior air space, bricks or concrete blocks and an outside air film. Heat will travel, by conduction, through all of these materials and its direction of travel will be dependent upon the season and whether it is warmer inside or outside of the house.

2. Radiation   Heat energy can be transmitted by heat waves or rays. This energy moves freely in space and has some unusual properties. When heat waves come in contact with a material, they may pass through the material as light. In this case, the material is not warmed to any degree. The rays may be reflected away from the material if it has a smooth, bright surface. Or they may be absorbed by the material if it has a dark or rough surface, Figure 3-9. When the material prevents the free passage of heat waves, it becomes hotter.

Radiant heat is always present to a degree and heat waves travel from one's body toward a cold window, from a fireplace to walls, furniture, etc. The direction of radiant heat, as before, is always from the warmer to the colder body. Radiation is not considered a major factor in air conditioning but provision is made in commercial *heat gain tables* for the types of surface on roofs.

3. Convection   Heat, particularly heated air, has another characteristic which is most useful in heating design. As air temperature increases, the volume of air expands. As it expands, it gets lighter in weight and rises. The cooler, heavier air settles in the bottom of the space. The warmer it gets, the more rapidly it will move and this rising of warm air is called *convection*. Thus, the air in a room is constantly moving, trying to equalize the temperature in the room. Baseboard radiation heating systems use this principle to distribute warm air throughout a room, Figure 3-10. Conversely, if warm air rises, cool air will fall, forming a natural convection *current* of constantly moving air.

All of these basic principles are used in a modern furnace. First, the combustion process releases heat energy which is confined in a metal heat exchanger. The walls of the heat exchanger will become hot, nearly as hot as the fire. Cooler air, passing along the outside surface of the heat exchanger, picks up this heat energy and becomes warmer. (An electric furnace eliminates this step and supplies heat energy directly to the air.) The heat exchanger is designed so that the air will pass over its entire surface, picking up a large amount of the heat energy available, Figure 3-11. Most of this transfer—fire to heat exchanger to air—is conduction, but some heat passes due to radiation.

As the air enters the space, it rises (due to convection and the force of the blower), mixes with the air in the space and warms it. It not only warms the room air (cooled by heat loss through windows, walls, etc.), but also replaces the air lost through the structure. Since the warm air does rise, supply registers are almost al-

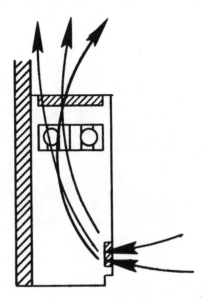

FIG. 3-10. Convection currents passing through a baseboard radiator.

FIG. 3-11. A furnace heat exchanger transmits heat by conduction and radiation.

ways placed in the floor and under windows. This location allows the warm air to pick up the cold air falling down from windows or other cool surfaces and and mix with it, Figure 3-12. This picking up of cold air, coming down from the windows, is called *entrainment* and helps eliminate drafts from the house.

Thus, in a ducted forced warm air heating system, air is used as a heat transfer medium to move heat energy from its source (the furnace burner) to the space where it maintains predetermined comfort conditions.

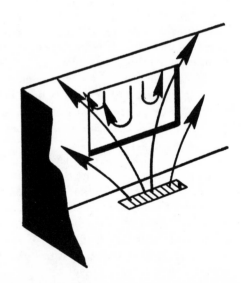

FIG. 3-12. Warm air from the register mixes with the cold air and *entrains* it.

CHIMNEYS AND FLUES

As previously mentioned, the products of complete combustion are carbon dioxide, water vapor and nitrogen. Incomplete combustion produces all these products plus carbon monoxide, aldehydes and soot. The flue products of complete combustion should be vented to the outside atmosphere because, though not dangerous, they are undesireable within the living space. In the case of incomplete combustion, they are dangerous and *must* be vented to the outside.

Therefore, proper venting of these gases is a most important part of any heating system. Most homes with gas or oil-fired appliances will have a chimney running from the lowest floor to the roof. Houses without a chimney require a flue and the same principles apply. The furnace should be located as close to the chimney as possible to minimize flue length. The furnace is connected to the chimney with a flue pipe, Figure 3-13.

If more than one appliance is vented into the chimney, the one with the largest flue should be the lowest, Figure 3-14. An alternate method is shown in Figure 3-15.

As air is heated, it expands and rises and an increase in heat will increase the velocity of this movement. This is the basic principle behind the operation of the flue.

A chimney is a confined space leading upwards to the outside air. If the temperature of the air surrounding the pipe is the same as the air within the pipe, then there will be no air movement. However, if the air inside the pipe is heated, it will expand, become lighter and rise. The warm-

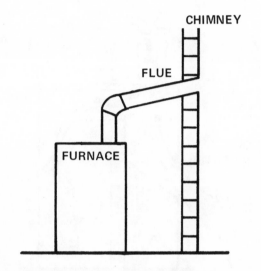

FIG. 3-13. The furnace is connected to the chimney with a flue.

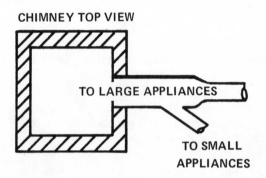

FIG. 3-15. Recommended methods for joining two flues.

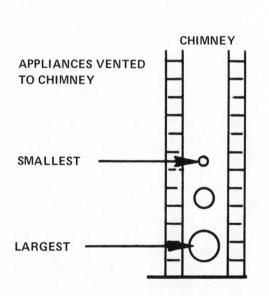

FIG. 3-14. The larger flue should always be below the smaller flue.

er it becomes, the faster it will rise. A steady, upward flow of warm air is said to create a good draft. As the warm air rises, it also creates a void which is filled by new air. If this air is also heated, it too will rise, continuing the flow for as long as there is a source of heat.

Obviously, a draft can only be maintained when the furnace is *on*. This works fine because a draft is needed to remove products of combustion, only during the combustion process.

Sizing of the chimney is important because the volume of flue gas that it can remove depends upon the temperature of the gas, plus the height and diameter of the chimney. If the chimney is too wide or too short, the gas loses its velocity and will not flow out of the chimney. If the diameter is too small, the resistance due to friction will reduce the draft. An extremely cold chimney exterior can hinder the draft and actually cause a downdraft until the inner surface heats up, due to the flue gases cooling off and slowing down before they reach the outside.

Another potential problem with a cold chimney is condensation. As noted under *Products of Combustion,* a considerable amount of water is produced during the combustion process. This moisture is carried up the flue and vented to the outside. A large stone or masonry chimney will not warm up quickly and can cause a problem of condensation within the chimney. Sulphur is normally present in flue gases. Combined with water vapor, sulfur produces sulphuric or sulphurous acid, which can eat the mortar out of the chimney and actually go through certain types of masonry.

Where it is necessary to vent through a large stone chimney, installation of a liner, with an insulating air space around it, is strongly recommended. Any liner should be installed with overlapping joints, so that any condensation will drain all the way down to the bottom of the chimney. The problem of condensation can sometimes be solved by putting in an oversized diverter at the furnace, so that more warm air from the furnace room is drawn in and up the chimney.

If condensate forms in the flue pipe, then insulating the pipe may help solve this problem.

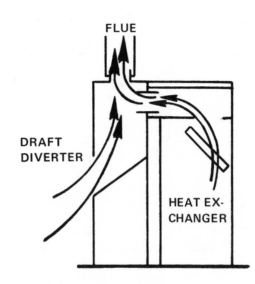

FIG. 3-16. Gas furnaces are equipped with a draft diverter.

VENTING A GAS FURNACE

About 20% of the total heat produced by the furnace is used to heat the flue gases so that they are warm enough to give good draft. The other 80% goes to heat the house. This is why a furnace's heating efficiency is rated at 80% of input. A gas furnace uses this principle to assure sufficient air to have complete combustion. By matching the heat exchanger air volume and burner temperature, a draft effect is produced, bringing the proper amount of combustion air into the compartment. This amount of air is independent of the chimney draft whose primary function is to get rid of the products of combustion.

Gas furnaces will also have a relief device called a draft hood or diverter built into the cabinet between the furnace and the flue, Figure 3-16. This prevents the chimney draft from drawing excess combustion air into the furnace, thereby lowering the temperature of the air flowing over the heat exchanger and reducing efficiency. The diverter is connected to the top inside of the heat exchanger where flue gases are collected from the combustion process. The bottom is open to the surrounding air. The flue running to the chimney connects to the top or side of the diverter.

In a properly operating gas furnace, the flue gas passes from the top of the heat exchanger, along the top of the diverter, then into the flue and up the chimney. The chimney draw is usually more than is necessary to remove just the flue gases so additional room air is also drawn in at the bottom and then up the chimney.

The chimney draft can be upset by several factors. There can be a restriction in the chimney which reduces the volume of gas the chimney can handle. There can be complete stoppage of the chimney where no gas can be vented outside. There is also the possibility of a downdraft due to the outside temperature or wind. Note particularly that the furnace can operate even if the flue or chimney is *completely* stopped up.

If any of these things happen, the flue gas has to have some place to go or else the back pressure will upset the combustion process in the furnace. The diverter provides this relief and allows the gases to vent into the room. This is *spillage.*

When this occurs, the flue gases are circulated into the house. In the case of complete combustion, this is not particularly critical as its effect will be to raise the humidity and circulate harmless carbon dioxide. If, however, combustion is not complete, carbon monoxide and aldehydes also will be circulated. These gases are extremely dangerous and can be fatal. *Anytime spillage is discovered, the cause must be found and corrected immediately.*

There is a simple and positive test for draft spillage. There will likely be a gas odor and the homeowner may have noticed high humidity. If the homeowner complains of running eyes, headaches or dizziness, incomplete combustion and spillage should be suspected. To test for spillage, light a match or use a lighted cigarette and place it about ½" below and in front of the diverter with the furnace operating, Figure 3-17. If the flame or smoke is drawn up the diverter, it is working normally. However, if the flame or smoke is blown back into the room, then there is spillage and the cause must be found and corrected.

NOTE If there is the appearance of spillage, wait a few minutes and recheck. Temporary spillage, caused by outside winds momentarily reversing the chimney draft, is normal and not a problem.

VENTING AN OIL FURNACE

Because an oil furnace is supplied with a combustion air blower, the removal of flue gases is somewhat different. Since an oil furnace is not dependent upon the furnace chimney effect for combustion air, no

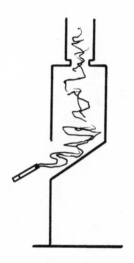

FIG. 3-17. Checking a draft diverter using cigarette smoke.

FIG. 3-18. Barometric damper for an oil furnace flue.

diverter is required. The oil furnace is piped directly to the chimney.

A device called a *barometric damper* is installed in the flue pipe to provide draft control. The damper is hinged in the middle so it can swing freely and has an adjustable weight, Figure 3-18. When the draft becomes excessive, the damper will swing open, drawing room air into the flue. The weight is normally set at approximately 0.04 inches of draft using a water gauge.

Combustion in an oil furnace is not normally upset by downdrafts because the combustion air blower has enough power to overcome them.

An oil furnace with improper combustion usually produces excessive carbon or soot rather than carbon monoxide. There is less water vapor produced than with a gas furnace, so a diverter or other combustion relief is not too important.

A good flue and venting system is important for proper operation of all furnaces. While the serviceman has no control of the original application, he should check the following points on both the flue and chimney as they can materially affect furnace operation.

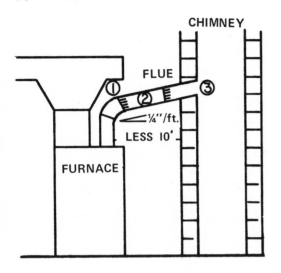

FIG. 3-19.

7. The pipe should be clear of all obstructions.
8. If there is poor draft or condensation in the pipe, insulating the pipe may help correct this condition.

NOTE   After a heavy snow, flue pipe outlets should be carefully checked to be sure that snow on the roof is not obstructing the flue.

FLUE TO CHIMNEY, Figure 3-19.
1. Joints must be tight and in direction of flow to offer minimum resistance. Fitting at chimney must be tight.
2. The flue pipe connecting the furnace and flue must be the same size as the furnace collar through its entire length.
3. The flue pipe should not extend beyond the inner face or liner of the chimney.
4. Horizontal flue pipes should be as short as possible and never more than 10 feet long to avoid friction losses and maintain flue gas temperature.
5. The flue pipe should be as straight as possible to avoid friction losses.
6. The pipe must have a pitch upwards of at least ¼" per foot of length in horizontal runs.

CHIMNEY, Figure 3-20
1. The top of the chimney should be at least 2' above the highest point on the roof to prevent downdrafts.
2. The chimney top must also be at least 3' higher than the point at which it passes through the roof.
3. The coping should not restrict the opening.
4. The chimney should be free from loose building materials, displaced bricks or other debris. Check by lowering a flashlight down the chimney.
5. Joists or other structural members should not protrude into chimney. Check with flashlight.
6. There should be no leakage between tiles or chimney face. This can be checked by starting a smoking fire and observing if leakage occurs.
7. The cleanout door should fit tightly.
8. There should be no opening between flues using a common chimney.

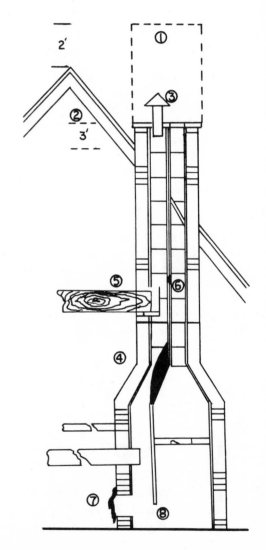

FIG. 3-20.

Another common cause of improper draft is a low or negative pressure at the furnace. If the pressure in the area is lower than atmospheric pressure, it can cause a downdraft which results in spillage or poor combustion.

The flue system requires an adequate supply of combustion air to work properly. Anything that reduces the normal air supply obtained from infiltration and circulation will in effect *starve* the furnace. The most common causes of this condition are:

Insufficient Furnace Room Air If air flow into the furnace room is restricted in any way or any part of the air entry to the furnace is blocked or covered up, the furnace will be *starved* for combustion air. Additional air openings, such as a louvered basement door or an outside air intake, will usually solve this problem.

Furnace Blower Compartment Door Is Off The suction of the blower can cause a negative pressure in the immediate area which can unbalance the flue system and result in spillage. The products of combustion are therefore circulated throughout the house. This door should be kept on when the furnace is operating.

Exhaust Fans or Vents Near the Furnace Clothes dryer vents, kitchen exhaust fans or window-mounted exhausts located near the furnace will all remove air from the immediate area. If this is the case, additional openings into the furnace area are needed and, again, possibly an outside air intake.

The serviceman should make certain, when running any checks or tests on the furnace, that he does this under normal operating conditions so the results he obtains will give him a true picture. Doors should be on the furnace, windows closed, basement door shut, exhaust fans running, etc. Even if an exhaust fan only runs intermittently, it should be on to give him results under the most adverse conditions. In this way, he can diagnose and repair the problems more accurately.

FURNACE APPLICATION

Furnaces can be applied in a variety of different ways, depending upon the type of house and its basic construction. These fundamental applications are independent of the energy source used by the furnace, and so the physical location of the furnace cabinet and its duct work will be the same for gas, oil and electric furnaces. Generally speaking, the choice of energy source will be made on the basis of which fuel provides the most value for the dollar, on the availability of various fuels in a given area, or on the basis of homeowner's preference for a fuel he likes and is comfortable with.

The furnace cabinet contains the blower, heat exchanger and all attendant controls to make them operate, again regardless of energy source. See Figure 3-21. The blower is always located between the return air inlet and the heat exchanger, and the filters are always located between the return air inlet and the blower. When an air conditioning system is installed, in conjunction with the furnace, the air conditioning coil will be located between the heat exchanger and the supply air plenum.

Furnaces are classified by the direction of air flow through the furnace. For

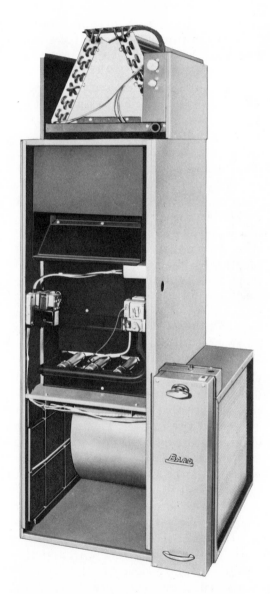

FIG. 3-21. Exposed view of a gas furnace showing blower, burners and heat exchanger. *Courtesy Bard Manufacturing Company.*

instance, if the return air enters the furnace at the bottom or side and flows up through the furnace and into the supply ductwork, this type of furnace is called an up-flo furnace, Figure 3-22. An up-flo furnace will normally be used in a basement where the supply air runs can be put up high in the basement ceiling, directly underneath the floor of the living space.

Another type of application is where the return air enters the top of the furnace and flows downward through the furnace. Here, the supply air plenum and ductwork are located at the bottom. A closet application in a single-story house will also take in return air at the top and distribute the supply air through duct running under the floor. This type of application could be used in a house without a basement that requires perimeter type distribution. This furnace is called a down-flo, reverse-flo or counter-flo, Figure 3-23.

A third type of basic application is where the air flows horizontally through the furnace, entering at one end and discharging at the other end. This is called a horizontal type furnace, Figure 3-24. The basic application for this furnace would be in a house with a crawl space. The furnace is located in the crawl space and the duct system runs under the floor, terminating under the windows. In a second type of application, this type of furnace is put into an attic space and the supply air either ducted down through the wall and discharged low, or ducted directly to a high side wall or through ceiling diffusers. This type of horizontal flow furnace, and/or combination heating and air conditioning system, can also be put outside, with the

FIG. 3-22.   Air flow in an up-flo furnace.

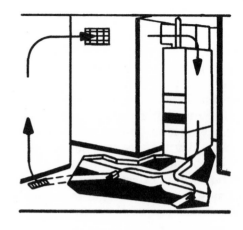

FIG. 3-23.   Air flow in a down-flo furnace.

FIG. 3-24.   Air flow in a horizontal type furnace.

supply air running through the wall, into the house duct system. This is quite common in mobile home applications.

In the conventional system, the supply air is carried by the main trunk duct, of-

ten called an extended plenum, which runs lengthwise through the house. Smaller ducts, running off the main duct, supply the various registers, either in the floor or on the wall, as in Figure 3-25. A supply air plenum can also be built as a box directly above or below the furnace with all the take-offs coming from a central point rather than the extended type plenum.

Air is returned to the furnace through a return air system using grilles located in central collecting areas, such as stairwells, or grilles cut into a sidewall or the floor. This air is then ducted back to the furnace. The simplest type duct or return air run can be made by placing a piece of sheet metal across the floor joist (called a panned joist), Figure 3-26, and using this cavity to return the air to the furnace. Many furnaces will have what is called a return air drop, which is merely a small cabinet designed to fit beside the furnace itself. Its purpose is to direct the return air from the top, where it is collected from the return air duct system or panned joist, to the bottom of the furnace. Many furnaces have two compartments: one for the heat exchanger, gas valve, burners, diverters, coil and electrical, and a matching cabinet for the blower, electrical make-up, humidifier and electronic air cleaner. This second cabinet can also act as a return air drop.

Most systems also require some additional outside fresh air to replace air lost through the walls and, in the case of gas or oil furnaces, to assure a constant supply of combustion air in the basement. A fresh air intake will be placed outside and connected into the return air system somewhere ahead of the furnace filters, Figure 3-27. This allows the fresh air to mix with the return air, raising its temperature, prior to passing over the heat exchanger.

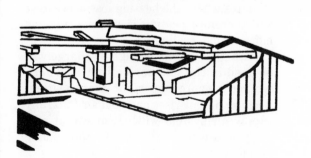

FIG. 3-25. Supply ducts may be installed in the floor (above) or the ceiling (below).

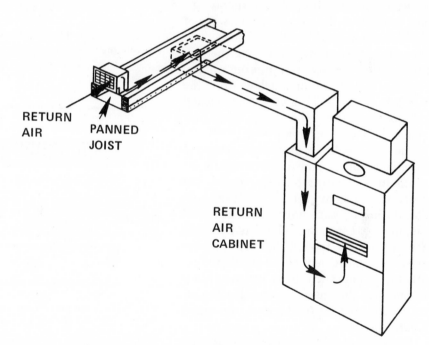

RETURN AIR

PANNED JOIST

RETURN AIR CABINET

FIG. 3-26. Typical return air system using a panned joist.

In colder climates, where there is a possibility of condensation and frost, the return air pipe is insulated. It will also be equipped with a manual damper to control the amount of fresh air allowed into the system.

Air is delivered to the individual rooms through a series of supply registers or diffusers, Figure 3-28. In colder climates, these are placed on the floor, directly under the windows, and about 18" from the outside wall. This placement will allow drapes to be closed and prevent other obstructions, close to the window, from interfering with the flow of supply air. The velocity of the air, as it comes out of the diffusers beneath the windows, will cause it to flow upwards, wiping the surface of the window and picking up any cold drafts that are coming down from the window surface. This process is called *entrainment*, since the warm supply air mixes with the cooler air coming off the window to create air at an even temperature. This mixed air will circulate in the room in a circular pattern, returning down to the floor. But, since it is thoroughly mixed with both warm and cool air, no draft is experienced as it moves across the floor toward the return air register. The movement of the air, therefore, allows warmed air to circulate freely throughout the space, eliminating the effect of the cold air falling to the floor and creating a draft.

This circulation will continue as long as the furnace blower is running. This is why it is suggested that the blower be allowed to run all the time, whether the furnace burners are on or not.

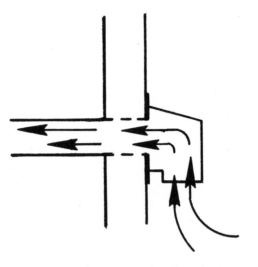

FIG. 3-27. A fresh air intake allows fresh air to mix with the return air.

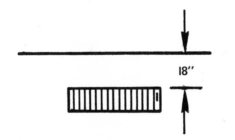

FIG. 3-28. Supply register.

FIG. 3-29. Dampers in run (above) and in the register (below) control air volume through the register.

Even though the supply air temperature, when the furnace is on, will usually be some 90° F above room air temperature, it will feel cool to the touch as it discharges from the diffuser because of its velocity. However, this effect will not cause any discomfort as long as the person is not sitting directly in the supply air-stream.

Air is controlled first by the size of the duct runs (calculated in the initial installation). There will also be a damper in the run, close to the furnace or the take-off from the plenum, which can control the volume of air through that particular run. In addition, there will be a damper at the register which also controls the volume of air. These two dampers therefore allow the balancing of the air volume in the system, Figure 3-29.

As a general rule, the location and type of furnace selected will be very much dependent on the house construction and whether it has a basement, crawl space or attic space available for the equipment. The supply duct system will be dependent upon the type of climate, with the perimeter floor system being used primarily in climates where the heating loss is greatest and therefore heating is as important or more so than cooling.

A high side wall or high discharge will be used in the southern climates where cooling is more important than heating. Because hot air rises and cold air falls, any supply air system will be somewhat of a compromise between heating and cooling efficiencies.

Questions

1. What is kinetic energy and why is it important?
2. What is temperature?
3. What is a Btu?
4. What three things must be present in order to achieve combustion?
5. Name three fuels commonly used in heating.
6. What are the basic elements present in all fuels?
7. What are the products of complete combustion?
8. How much air is required for complete combustion?
9. What is incomplete combustion? Give an example.
10. What are the products of incomplete combustion?
11. What is a yellow flame and how does it occur?
12. What is a blue flame and how does it occur?
13. What is the significance of primary air and how does it affect the flame?
14. Name three situations where blue flame will have incomplete combustion.
15. How does heat flow?
16. Name three methods of heat transfer.
17. Why is a chimney and flue important?
18. Describe how more than one flue is attached to the chimney.
19. Name two major problems with a cold chimney.
20. How much heat produced by the furnace is used to heat the flue gases?
21. Why is this important?
22. What is a draft diverter and why is it used?
23. How can chimney draft be upset?
24. What happens if it is?
25. Describe a simple test for spillage.
26. How is draft control accomplished in an oil furnace?
27. Name six points to cover when checking the flue connection to the chimney.
28. Name six points to cover when checking chimney construction.
29. Name three things that can cause *starving* of the furnace.
30. How are furnaces classified? Name them.
31. What other connections are made to the furnace?
32. How is the air delivered to the individual rooms?
33. Where are the registers located and why?
34. How is the amount of air delivered to the room controlled?

Chapter **4**

HEATING COMPONENTS

Many heating components are common to all types of furnaces regardless of the energy source. These components—blowers, filters and cabinets—will be covered first, then the components which vary according to the type of furnace. There will be variations in some of these components depending upon different manufacturers' applications, but the principles remain the same. Electrical characteristics will be covered later. It will suffice for now to be able to recognize the basic types of residential equipment.

CABINETS

Furnace components are located and mounted in a sheet metal cabinet which forms an airtight enclosure. A collar is provided for the attachment of the supply air plenum. For an up-flo furnace, this will be on top. Since the manufacturer does not know which side will be used for return air at the time the furnace is built, the return air opening is left intact, but usually indicated by knockout panels on each side, Figure 4-1. This opening is then cut out on the jobsite. Some furnaces also have a knockout on the bottom for specific applications like a mobile home or apartment where the return air comes in below the furnace. Knockouts are also provided for electrical connections and fuel piping. Size of the cabinets will vary considerably between different manufacturers, because the size is dependent upon the space required for the heat exchangers and burners plus space to move the required amount of air. Where height is the major consideration, as in a low-boy furnace designed for basements with low ceilings,

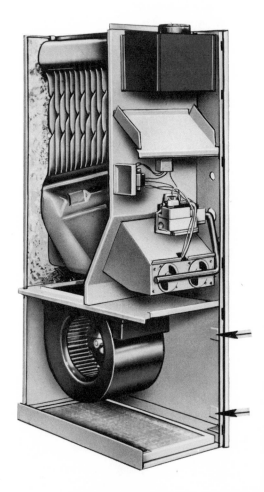

FIG. 4-1. Knockouts for a return air opening. *Courtesy Rheem Mfg. Co.*

FIG. 4-2. (Left) Up-flo furnace. *Courtesy WeatherKing, Inc.* In low-boy applications of the up-flo principle, two cabinets are placed back to back. *Courtesy Century Engineering Corp.*

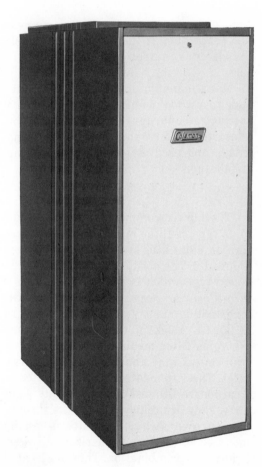

two cabinets will be placed-back-to back. Where height is not a factor, all components are combined in the one cabinet, Figure 4-2.

Furnaces require access panels or doors to get at the burners, blower compartment, and other operating components. These may snap on or off or, in some cases, latching handles are provided. The doors may have louvers to allow com-

FIG. 4-3. An electric furnace requires no openings for combustion air. *Courtesy The Coleman Co., Inc.*

bustion and vent air to enter or have just an open area. Because there is no combustion, and thus no need for combustion air or air for venting, an electric furnace can have solid panels or doors, Figure 4-3. Electrical *makeup boxes*, for line and low voltage, will be located in one of these compartments, usually the blower compartment.

A gas or oil furnace will have an oval flue connection at the top of the cabinet. This is not required for an electric furnace.

Many manufacturers will bolt the furnace to a wooden shipping frame with a bolt at each corner. These bolts can be reinserted, after removing the shipping frame, and used as leveling bolts during installation.

FILTERS

The blower compartment will contain some type of filter media between the return air side of the system and the blower, Figure 4-4. The most common type filter is a *slab* type which is available either as a *throwaway* or *permanent* type. Slab filters come in many standard sizes and are usually 1" thick. A throwaway type has Fiberglas filter media held in place by a thin metal face and back having large holes. These elements are then mounted in a cardboard frame. When these filters get dirty, they are removed from the furnace and thrown away. The filters can be mounted in the furnace horizontally, vertically or in combination. They are located near the return air inlet and ahead of the blower, Figure 4-5. A rack or a rail is provided for easy insertion and removal.

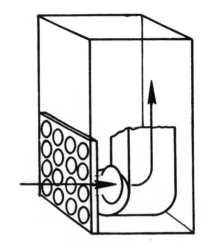

FIG. 4-4.   Filters are inserted in the return air stream ahead of the blower.

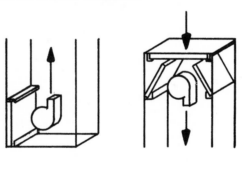
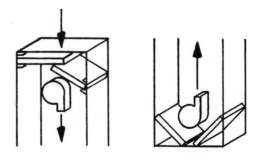

FIG. 4-5.   Filters can be mounted horizontally, vertically or in combination.

The *permanent* type can be either polyurethane or a metal fiber mesh. The polyurethane filter media mounts in a frame, which is usually made of wire, with clips to hold the media in place. The media in a metal fiber filter is not removable from its frame, Figure 4-6. Either type can be removed from the furnace and cleaned. They are then reinstalled in the furnace.

NOTE   On a down-flo furnace, it may be necessary to remove the flue pipe and side panels to get at the filter.

CLEANING SLAB FILTERS

To clean a polyurethane filter media:
1. Remove media from the rack.
2. Wash the media in water, using a special cleaner.
3. Squeeze the water from the media.
4. Replace the media in frame.
5. Oil the surface of the media and reinstall in the furnace.

FIG. 4-6.   Typical wire mesh filter. *Courtesy Research Products Corp.*

Metal fiber type filters can be easily cleaned with high pressure air or water as found in a do-it-yourself car wash.

Another filter is a *wraparound* type which fits on a rack in the blower compartment. This rack slides in or out on rails in the cabinet. The media can either be disposable or permanent. The permanent type can be cleaned in the same manner as slab filters.

Filters have varying degrees of efficiency—that is, their ability to remove particles from the airstream. As they fill up with collected particles, they become more efficient (they remove smaller sized particles) but the restricted air flow makes the furnace less efficient and can cause it to shut off on limit.

BLOWERS

The prime air mover, and one of the most important components in the heating system, is the blower, Figure 4-7. The blower has an 8 to 10" diameter wheel or scroll mounted in a metal enclosure or housing with an opening offset to one side. When the blower rotates, air is taken in at the side and pushed out the opening at the top under pressure. There are two basic types of blowers, depending upon how the motor is mounted: belt drive and direct drive.

In a belt drive blower, the wheel or blower is mounted on a shaft set in bearings on each side. A large grooved pulley is mounted on the end of this shaft. The motor is mounted on a platform or cradle, and both motor and mounting hardware are cushioned with resilient rubber to minimize noise and vibration. The motor

mounting frame is designed so that the motor can be moved left and right or up and down. This is necessary so that the pulley on the motor can be exactly aligned with the blower pulley. The motor is usually attached to the cradle with two adjustable-tension metal straps, one on either side.

The blower belt should be examined for cracks, splits, ragged edges and uneven wear. If any of these are present, the belt

should be replaced. Be sure to note the proper belt number and size.

SERVICING THE BLOWER ASSEMBLY

To Remove the Blower Belt:
1. Turn the furnace off.
2. Loosen the motor mount screw.
3. Move the blower motor up to release belt tension.
4. Remove the belt from both pulleys.

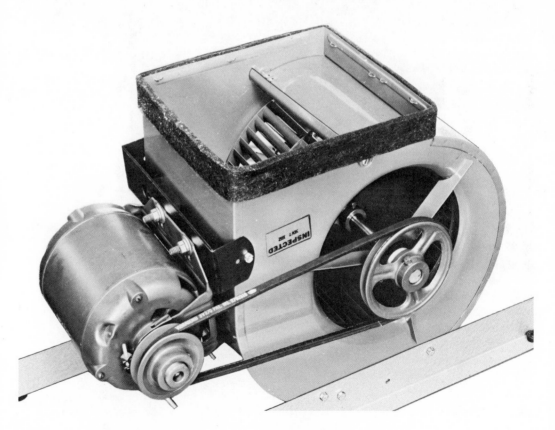

FIG. 4-7.   Belt-driven blower. *Courtesy Century Engineering Corp.*

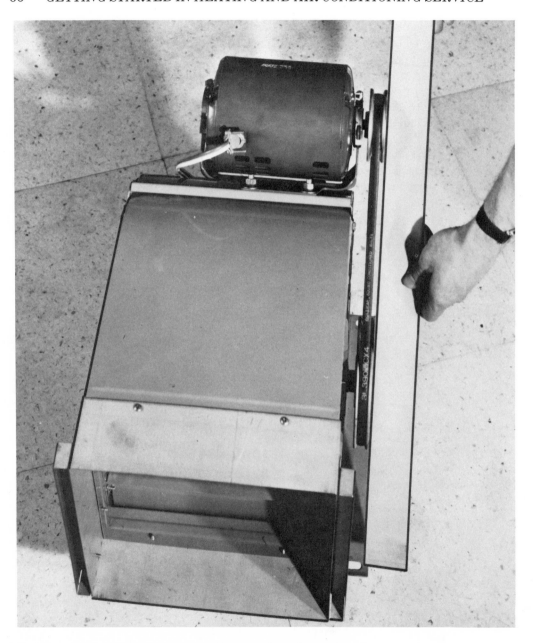

FIG. 4-8.   Checking pulley alignment. *Courtesy The Coleman Co., Inc.*

Checking the Blower and Motor Bearings (Belt Off)

1. Check the set screw on the blower pulley for tightness.
2. Try to move the blower shaft up and down, back and forth. The shaft should not move in any direction. If it does, the bearings are bad and need to be replaced.
3. Try to push the blower shaft in and then pull it out. If there is greater than ⅛" play, loosen the set screw on the thrust collar and move collar closer to the bearing and retighten the set screw.
4. Try to move the motor pulley up and down, back and forth. There should be no movement. If there is, the motor bearings are bad and need to be replaced.
5. Push the motor shaft in and out. There should be about ⅛" end play.

Checking Pulley and Drive Alignment

1. Place a straight edge along the side of the pulleys or a straight ¼" diameter steel rod in the grooves of both pulleys, Figure 4-8.
2. The rod should be absolutely straight in the grooves of the pulleys. If not, loosen shaft set screw on motor pulley and realign.
3. Tighten set screws and recheck with rod.
4. Recheck all set screws for tightness. Check motor bracket screws for tightness.

Lubricating Motor and Blower Bearings

1. If bearings have no grease cups or fittings, they are presumed to be permanently lubricated and need no further service.
2. If bearings have grease cups, turn down the cup about 1 turn yearly. Refill when the cups are turned all the way down. Follow manufacturer's instructions as to type of grease.
3. If bearings have grease fittings or plugs, remove the plug from the bearing, add a 5 lb relief fitting and lubricate with a No. 2 consistency, neutral mineral grease.
4. Before lubricating motor bearings, check for manufacturer's instructions on the nameplate. If no instructions are available and the motor has oil cups or holes, add a few drops of SAE No. 10 nondetergent oil twice a year.

NOTE   Do not overlubricate, as excessive oil can damage the starting switch, spill and collect dust or damage belt.

Checking Belt Tension

1. Press down on belt with finger.
2. Belt should deflect about 1" when moderate finger pressure is applied to the top of the belt halfway between the pulleys, Figure 4-9. If not, loosen the mounting bracket screw and readjust. Retighten screw.

FIG. 4-9.   Checking belt tension. *Courtesy The Coleman Co., Inc.*

Most belt drive blowers have an adjustable motor pulley which is used to change the blower speed. This pulley is sometimes needed to adjust air volume during the initial heating installation and to increase the air volume when a changeover to cooling is made.

Adjustable pulleys have an Allen setscrew in the outer flange which, when loosened, allows the flange to be screwed in or out, Figure 4-10. Some pulleys have two flat areas in the threads that are 180° apart, so an adjustment can be made every ½ turn. Others have only one flat area, so speed adjustments can only be made for one complete turn. Be sure to locate the setscrew over the flat area of the shaft, since another location can strip the threads. Rotating the outer flange clockwise, increases the inside diameter of the pulley and makes the blower run faster. Rotating the flange counterclockwise decreases the inside diameter of the pulley and reduces blower speed.

Since pulleys, motors and drives are matched to the furnace air requirements, the range of adjustment is usually capable of supplying the necessary air for most normal applications anywhere in the country. If more air is required, it is possible to change the entire pulley to the next larger size. Caution should be exercised in changing pulleys because, if a great deal of additional air is required, it may also be necessary to change the blower motor to the next larger size. If the motor is too small, it will not supply the air required and will be overloaded. How to check this will be shown later.

A direct drive blower is one where the

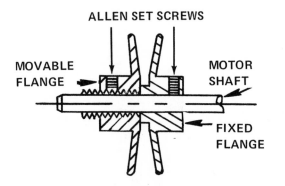

FIG. 4-10. Adjustable pulley.

FIG. 4-11. Direct drive blower. *Courtesy Century Engineering Corp.*

blower motor is mounted inside the blower scroll and the drive shaft is connected directly to the wheel, Figure 4-11. Therefore the blower will run at the same speed

as the motor. A direct drive blower is closely matched to the furnace and its intended application. If there is a need to change blower speeds, it must be done electrically. This will be discussed in the next section.

Maintenance checks of setscrews and bearings, as well as lubrication, follow the same procedures outlined under belt drive blowers.

BLOWER MOTORS

There are several different types of motors used to drive the furnace blower, each selected according to the running and starting loads imposed by the application. These motors are the:

Split-phase motor   A split-phase motor develops starting torque by having the starting and running windings out of phase with each other, Figure 4-12. When

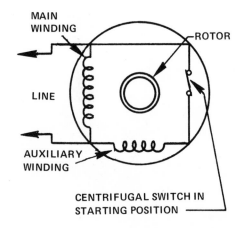

FIG. 4-12. Schematic of a split-phase motor.

the motor starts, both windings are in the circuit but, when the motor comes up to speed, a centrifugal switch opens, taking the start winding out of the circuit. This motor has relatively low torque which means it can only be used on applications with light starting loads. It is used on belt drive applications in 1/3 hp and smaller sizes. This is usually a 4-pole motor which operates at 1750 rpm.

Capacitor-start motor   A capacitor start motor has a starting capacitor mounted on the motor and internally wired into the start winding, Figure 4-13. This additional assistance to the start winding gives the motor considerably more starting torque so that it can start against greater loads than a split-phase motor. When the motor comes up to speed, a centrifugal switch opens, taking the start winding out of the circuit. A capacitor-start motor is used on belt drive applications in 1/3 hp and larger sizes.

Shaded-pole motor   A shaded-pole motor gets its starting torque from loops of copper wound around each pole which puts the windings enough out of phase to start under relatively light loads. It has the lowest starting torque of all motors described here and its application is limited to the smaller residential furnaces. The shaded pole motor is used on direct drive blowers in 1/10 to 1/6 hp sizes. It is a 6-pole motor operating at 1050 rpm.

Permanent split-capacitor (PSC) The permanent split-capacitor or PSC motor is a single-phase induction motor designed to use a single, fixed capacitor for both starting and running duty, Figure 4-14. The capacitor has two functions:

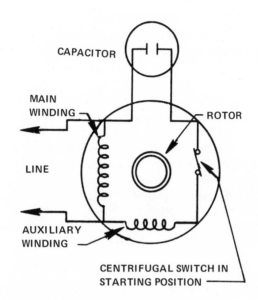

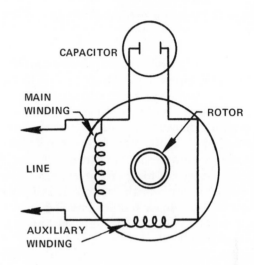

FIG. 4-13.   Schematic and exterior view of a capacitor-start motor. *Courtesy Century Electric Co.*

FIG. 4-14.   Schematic and exterior view of a permanent split-capacitor (PSC) motor. *Courtesy Century Electric Co.*

1. To help provide the extra starting torque needed when the motor starts.
2. To help improve the efficiency of the motor while running.

This capacitor is permanently connected in the circuit and splits the phase of the current in the auxiliary (starting or phase) winding with respect to the main (run) winding, hence the name, permanent split capacitor. Thus, a PSC motor has more starting torque than a shaded-pole motor and better running efficiency since the capacitor stays in the circuit. A PSC motor is also a 6-pole motor operating at 1050 rpm and is used on direct drive applications from 1/8 to 3/4 hp.

AMPERAGE DRAW

The amperage drawn by any motor is an indication of its operating condition. This reading should be taken and recorded during every service call. If the amperage is greater than the nameplate rating, it indicates that the motor is overloaded. This can be caused by restrictions in the air movement such as dirty filters. If the amperage is less, the motor is not working to full capacity.

MOTOR SPEED CONTROL

Direct drive motors are closely matched to the application but, just as in belt drive applications, blower speeds must sometimes be changed and thus the amount of air delivered to the space. The most common occasion for changing speed is in switching from heating to cooling. Speed can be changed electrically by using a speed control or series reactor. Since the speed of any motor is a function of voltage and load, varying the voltage, while holding the load more or less constant, will change the speed of the motor. Voltage can be varied with a series reactor or speed control. A reactor consists of an iron core wrapped with many turns of wire, similar to one side of a transformer. Usually 3 or 4 wire leads are tapped into the reactor at various points, Figure 4-15. If the motor is connected to the first lead of a speed control, the motor will run at high speed, because it is being supplied with full line voltage. If the motor is connected to the second lead, the motor will run at a slower speed because some of the voltage is used up in pushing the current through the first group of turns in the reactor. Therefore, the motor receives less voltage and runs slower. If the motor is connected to the third lead, the motor will run at its slowest speed because of the additional voltage used up in the reactor. Reactors should not allow a motor to run below 550 rpm, since at this point the lubrication of the motor bearings will be reduced to the point where the bearings will fail.

The taps are normally color-coded and then identified on the furnace manufacturer's wiring diagram. Unused taps should be taped separately and located out of the way to prevent any possibility of a short circuit. If the taps are not identified, a voltage reading across them will help determine the speed—the higher the voltage, the greater the speed. The series reactor can be used with voltage sensitive direct drive motors and is either supplied on the original installation or can be added to existing applications which have either shaded-pole or split-phase motors.

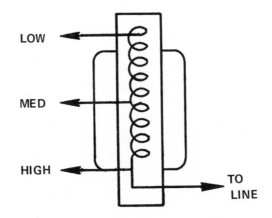

FIG. 4-15. Series reactor used to vary motor speed.

TAP-WOUND MOTORS

Motor manufacturers have applied this same principle in another way to provide multispeed motors for direct drive applications. This is a tap-wound motor with 3, 4, or 5 speed choices. Both shaded pole and PSC motors are available with this feature, Figure 4-16. By tapping into the motor windings and bringing these leads outside, many speed choices can be easily made. Just as with a speed control, the greater the length of winding used, the slower will be the motor speed. Leads are color-coded according to the speed they will produce. Most manufacturers will color-code as follows:

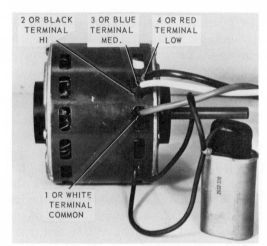

FIG. 4-16. Tap-wound, 3-speed motor. Courtesy The Coleman Co., Inc.

FIG. 4-17. The fan control is located in the heat exchanger. Courtesy Cam-Stat, Inc.

Color	5 Speed	4 Speed	3 Speed
Orange	Common	Common	Common
Purple	Capacitor	Capacitor	Capacitor
Black	High	High	High
Brown	Med. High	Med. High	
Blue	Medium		Medium
Yellow	Med. Lo	Med. Lo	
Red	Low	Low	Low

NOTE   Not all manufacturers have the same color-coding and this should be checked on the wiring diagrams before connecting.

As in a speed control, the unused leads should be taped separately to avoid shorts. Some leads have spade terminals which plug into a terminal box inside the furnace vestibule. This box may have insulated dummy terminals for the unused leads. The cooling relay may also be capable of selecting one of two motor speeds. This allows automatic selection, at the thermostat, of the proper speed for either heating or cooling. One initial selection of speeds is all that is required to have one speed for cooling and a slower speed for heating. This will be explained under *Cooling Components*.

FAN CONTROL

The fan control is included in all types of forced air furnaces, up-flo, down-flo, or horizontal. The fan control is located in the heat exchanger, about 1/3 of the way down from the top, Figure 4-17. The fan control has two major functions:

1. It cycles the blower on when there is a call for heat and turns it off when the thermostat is satisfied.
2. It provides a *time delay* before bringing the blower on and turning it off.

The reason for this time delay is to allow enough time for the furnace to warm up after a call for heat. When the air is delivered to the space, it will be up to proper temperature, avoiding drafts. When the furnace shuts off, additional air movement is desired so that the heat exchanger will cool down. This prolongs the life of a heat exchanger.

The fan control is in the line voltage circuit, in series with the blower motor, and contains a set of normally open contacts. The bimetallic element in the control is inserted into the heat exchanger so that it will be in the leaving airstream and directly register the temperature of the supply air. There are two types of fan controls: one in which both the *on* and *off* temperatures are adjustable and a second type in which only the *off* temperature is adjustable, with a fixed temperature differential for the *on* cycle.

In order to allow the blower to come on as quickly as possible, and direct the heated air to the space rather than to the flue, the set point for bringing the blower on is usually between 90 and 110° . The difference between the point where the blower comes on and the point when the blower turns off (which is called the differential) will run between 15 and 35° . In a fixed differential fan control, it will usually be 25° . This means that, if the blower is brought on when the air leaving the heat exchanger reaches 110° , the blower will shut off when the air cools down to about 85° . These set points can be changed to suit the preference of the homeowner and the lower the *on* set point is, without causing drafts in the room, the more efficient the furnace operation will be. With thermostats that do not have a fan switch for continuous blower operation, it is possible to set the fan control below the normal return air temperature and thus keep the blower operating even when the burners are off.

The calibration of the fan control can be checked by inserting a thermometer in the furnace plenum and noting the temperature at which the furnace comes on and also when it turns off. This reading will not be accurate because of the lag in the thermometer reading, but it will give a reasonable indication.

DOWN-FLO AND HORIZONTAL UNITS

Because heat rises, the fan control in an up-flo furnace will respond correctly under all conditions. However, in a down-flo furnace, the burners can come on, warm up the heat exchanger (the top gets hottest first) and raise the temperature of the fan bimetal control up to the *make* point. However, when the blower comes on, it forces cooler return air over the fan control. This cooler air could reduce the temperature to the *break* point, stopping the blower. This will cause cycling of the blower under perfectly normal conditions.

When the burners go off, it is desirable to keep the blower on long enough for the heat exchanger to cool down. Again, the return air will be at a temperature below the setting of the fan control and can *fool* the control in a down-flo furnace into thinking the heat exchanger has cooled off. As a result, it will turn off the blower too soon. Once the blower is off, the residual heat in the heat exchanger will rise, closing the fan switch and bringing the blower back on.

To avoid this, a special fan control is used on down-flo furnaces. This control includes an electric resistance or warp switch wired in parallel with the control circuit. On a call for heat, the warp switch will heat up before the bimetallic element in the furnace and close the fan control switch, starting the blower. This will take about 60 seconds and can occur without the burner coming on at all. As the supply air comes up to temperature, the bimetal in the furnace, acting through a push rod, will hold the contacts in. When the thermostat is satisfied, it opens the circuit and

the resistance will cool off rather rapidly. However, the residual heat in the furnace will cause the furnace bimetal to keep the blower on for an additional period (about 1½ minutes) before breaking the fan control contacts.

LIMIT CONTROLS

All furnaces are equipped with a safety device designed to prevent the heat exchanger from becoming overheated. This can occur if the blower motor fails or the air passing across the heat exchanger is restricted in any way. When a heat exchanger is overheated for any period of time, holes can develop, spilling the products of combustion into the supply air. This is not only a health hazard but can be a fire hazard, as well. A limit control, Fig-

FIG. 4-18. The limit control prevents the heat exchanger from becoming overheated. *Courtesy Cam-Stat, Inc.*

ure 4-18, is used to protect against this situation.

NOTE   This section applies to the limit controls used on gas and oil furnaces only. The limit for an electric furnace is quite a bit different and will be discussed in the chapter, *Electric Heating.*

The limit control is usually in the low voltage circuit for gas furnaces and in the line voltage circuit for oil, and has a set of normally closed contacts. It can be physically combined with the fan control, Figure 4-19, or it can be mounted separately. The limit control senses the temperature passing over the heat exchanger with a bimetallic element placed in the airstream. Limit controls are usually nonadjustable and have a factory-set break point at 200° F. When the limit opens, it breaks the control circuit to the thermostat, shutting off the furnace. A limit control will reset automatically when the furnace cools to about 175° F so there is a built-in differential, between make and break points, of about 25°

SECONDARY LIMIT
    To assure accurate sensing of all temperature conditions, a secondary limit is required for gas and oil down-flo and horizontal furnaces. (It is not necessary for an electric furnace, as will be discussed later.) The primary limit control, located in the lower part of the cabinet, will monitor for overfire or restriction of air over the heat exchanger. The secondary limit is wired in series and located above the heat exchanger. This limit will sense heat ris-

ing from the heat exchanger if the blower fails and open the contacts. On many oil furnaces, the secondary will be a SPDT switch, one leg of which will bypass the fan control and try to bring on the blower to cool off the heat exchanger and at the same time break the control circuit, shutting off the burner.

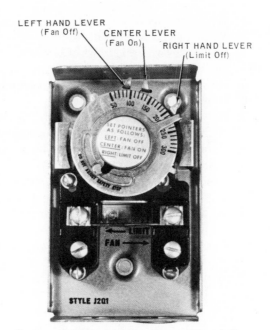

Courtesy White-Rodgers Div., Emerson Electric Co.

FIG. 4-19.   Three types of combination fan and limit controls.

Courtesy Honeywell Inc.

Courtesy Cam-Stat, Inc.

Questions

1. Where are the filters located in a furnace?
2. Name two basic types of filters.
3. Name the two basic types of blowers.
4. How do you check blower and motor bearings?
5. How do you check pulley and drive alignment?
6. How do you check belt tension?
7. How is the amount of air delivered by the blower changed in a belt drive blower?
8. Name four different types of blower motors and their application.
9. What is the speed controller and where is it used?
10. What is a tap-wound motor?
11. Describe the principle on which a tap-wound motor changes speed.
12. What is a fan control and why is it used?
13. What circuit contains the fan control?
14. What is a normal differential in a fan control?
15. Describe the fan control of a down-flo furnace.
16. What is a limit control?
17. Why is it in the circuit and which circuit is it in?

Chapter **5**
GAS HEATING

The heat exchanger in a gas furnace is often called a *clamshell* because it is formed from two pieces of metal stamped into left and right halves, Figure 5-1. These are then welded together along a common vertical seam. The metal can be cold-rolled steel or aluminized steel and may have a special ceramic coating to resist rust and acid deterioration.

The shape of the heat exchanger is often complex because it has several important functions. It must first provide room for the burners and allow them to supply steady, even heat over a maximum surface area inside the heat exchanger. Heat exchangers may have a reduced cross-sectional area, part way through the passage, to maintain the velocity of the air as it rises. Some heat exchangers will have baffles in the upper portion or, by their shape, force a zig-zag path. This is to assure that the maximum amount of heat, generated by the burners, will be transferred to the walls of the heat exchanger. Some heat exchangers will have dimples or indentations which serve still another purpose, that is, to give additional structural strength to the walls. This prevents cracking and popping noises *(oil canning)* during expansion and contraction of the metal.

At the top, the heat exchanger will have an exit for the products of combustion to flow into the diverter and flue. This area will be somewhat larger than the middle part of the heat exchanger.

The heat exchanger must be shaped so that the air passing over the outside surface will have the maximum wiping effect for the most efficient heat transfer. Air

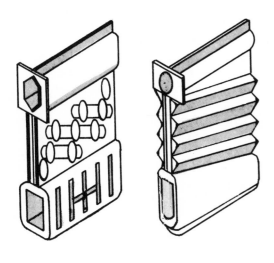

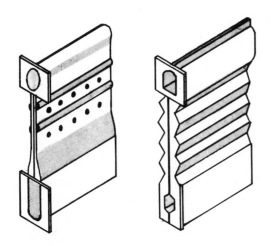

FIG. 5-1. Typical *clamshell* heat exchanger designs.

must pass over all the surfaces of the heat exchanger in sufficient quantity to remove the heat and avoid *hot spots* on the heat exchanger. If the air does not effectively remove heat from the entire surface, the *hot spot* will eventually burn through the heat exchanger, causing a failure. Several examples of heat exchanger design are illustrated in Figure 5-1.

Depending upon its effectiveness in achieving complete heat exchange and air wiping, each clamshell will have an output of between 20,000 and 50,000 Btu. Using more than one clamshell will increase the output rating of the furnace. For instance, a unit with two clamshells may be rated at 40,000 to 100,000 Btu, a three clamshell unit at 60,000 to 150,000 Btu, and so on. See Figure 5-2.

LEAK TESTING

A simple method for checking for leaks in residential furnace combustion chambers has been tested by the Detroit Commission on Buildings and Safety. First, a common table salt solution is sprayed into the combustion chamber, using a plastic spray bottle. Then the heated air is tested with a small butane torch. If any of the salt vapor from the combustion chamber has mixed with heated air (indicating a leak in the heat exchanger), the color of the butane flame will turn from blue to yellow. This is a very simple, quick, and economical check for defective heat exchangers that works equally well with gas or oil furnaces.

FIG. 5-2. Heat exchanger composed of three clamshells. *Courtesy Bard Mfg. Co.*

MAIN BURNERS

Each heat exchanger in the furnace will have an individual gas burner which fits into the opening at the bottom of the clamshell. Therefore, there will be as many burners as there are clamshells, the number varying with the Btu output of the furnace. Figure 5-3 shows a complete burner assembly.

Burners are designed to give a constant, even flame at maximum combustion efficiency. They must also be protected from lint and dirt since these are the primary causes of poor flame. They must supply sufficient primary and secondary air for efficient combustion and have a means of regulating the amount of primary air. Usually, the primary air is adjusted by means of a shutter or slide so that the quantity can be controlled. Primary air, as discussed under Combustion, is brought into the burner tube and mixed with the gas prior to ignition. Secondary air is drawn from outside of the burner and joins with the burning gases at or near the burner ports, Figure 5-4.

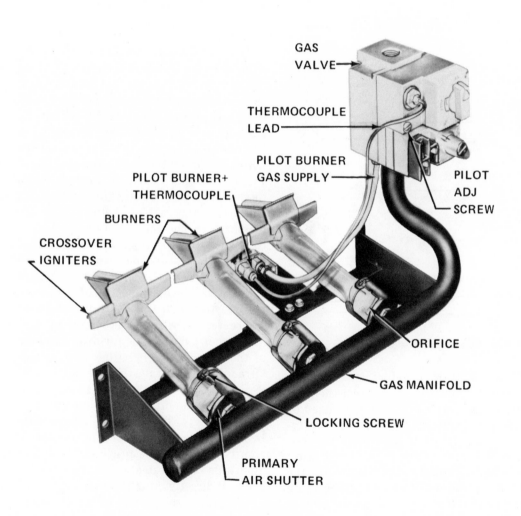

FIG. 5-3.   Complete burner assembly. *Courtesy Bard Mfg. Co.*

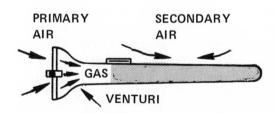

FIG. 5-4.   Supplying primary and secondary air to the burner.

In well-designed burners, the secondary air reaches the burner ports at an even rate, to provide a good gas-air mixture. Secondary air is brought to the burner by draft created in the heat exchanger, after the heat exchanger becomes hot.

Primary air enters the burner at the instant that gas flow begins, due to the suction created by the gas flow into the tube. Increasing the amount of primary air that is entrained with the gas reduces the dependence on secondary air and improves the performance of the burner.

If a burner has insufficient primary air, the flame tends to smother when it comes on in a cold heater. The flame will lift off the burner or puff out the front, which often extinguishes the pilot.

Too much primary air can produce a loud popping or *extinction* noise when the burners turn off. After the burner turns off, there is still a slight draft which will suck additional air into the burner. This mixes with the remaining gas and ignites at once in the burner, causing an objectionable noise and a puff which is strong enough to extinguish the pilot.

The amount of gas flowing into the burner is determined by the size of the gas orifice and the manifold pressure. The orifice is a small nozzle with a drilled hole to control the flow of gas. It is attached to the gas manifold and fits in a small hole in the center of the burner, Figure 5-5. The larger the hole, the more gas will be allowed to flow. The size of the orifice is determined by the specific gravity of the gas (usually about .60), the furnace input per orifice (about 25,000 Btu) and the heat value of the gas about 1,000 Btu/cu ft for

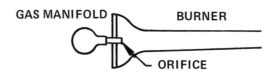

FIG. 5-5. Gas supply is controlled by the burner orifice.

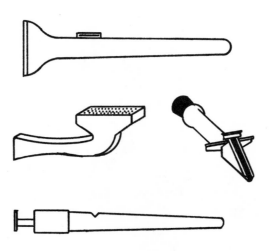

FIG. 5-6. Several burner designs.

natural gas. If these factors vary from the average figures in a given area, then the orifice size may have to be increased or decreased. Many manufacturers have tables or calculators to determine the new value. Orifice sizes are given as *drill* sizes. The smaller the drill size, the greater the Btu output. A natural gas furnace using 1,000 Btu/cu ft gas, with a specific gravity of .6 and a 25,000 Btu imput burner, would use a No. 38 drill size. If it has a 30,000 Btu input burner, it would use a No. 35 drill.

When the installation is at an elevation of 2000 feet or more above sea level, orifices must be changed. For operation at elevations above 2000 feet, input ratings should be reduced by 4% for each 1000 feet above sea level, regardless of the gas being used.

There are several basic burner designs, Figure 5-6. The most common is an elongated steel tube (some are cast iron) which has a throat or narrow portion (venturi) close to the entrance of the gas and a wider mouth, for primary air, to which the gas supply is attached. Gas is supplied to the end of this tube through an orifice which is connected to the gas manifold. The gas supply is fed to the end of the burner and, as it enters the venturi, its velocity increases. This helps pull in the primary air. The amount of primary air is controlled by either a shutter at the end of the burner or a rotating sleeve close to the throat, Figure 5-7. The gas-air mixture thus travels the entire length of the venturi tube to reach the combustion ports where it mixes with the secondary air.

Burner ports can be formed at the top of the burner tube by either an insert or an additional metal strip (usually stainless steel) between the open space. These can be formed either as ribbons or slotted crossways. The insert is held in place by spot welds or rivets so that it will constantly maintain the predetermined port openings. Other designs have holes or slots cut directly into the burner.

Several burners must be ignited quietly and simultaneously, with no delay even under low gas pressure conditions. Ignition is accomplished by a standing pilot

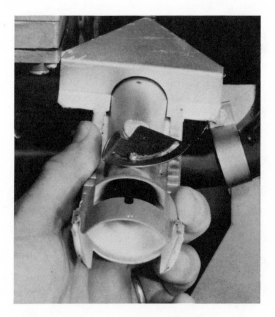

FIG. 5-7. Primary air is controlled by an adjustable shutter.

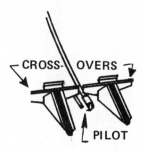

FIG. 5-8. Crossover igniter lights all burners simultaneously.

located close to the center of the furnace. A metal strip (or crossover igniter) which has a continuous open slot through it, is attached to each burner, Figure 5-8. On a call for heat, each burner will supply some gas-air mixture to the crossover. The pilot ignites the gas, bringing all burners on

FIG. 5-9. Normal pilot flame fans out in two directions.

simultaneously. Some crossover igniters will include a small baffle or air scoop to divert part of the gas-air mixture down toward the main burner.

Burners are held in place in the heat exchanger either by a small slotted piece of metal near the crossover igniter or by fitting them into a contour at the other end of the heat exchanger.

MAINTENANCE OF BURNERS

Proper burner flame can be observed by igniting the burners and visually checking. There should be a fairly soft blue flame with no yellow color. If the yellow flame is observed, indicating unburned carbon, then the primary air must be increased by opening the shutter or sleeve. The shutter or sleeve will have a small set screw to maintain the set position. This screw can be loosened and the air opening increased. The shutter is closed down until a yellow flame appears, then opened just to the point where the yellow flame disappears. The shutter should be set at this point and the set screw tightened.

In order to have the required blue flame, it is necessary to have approximately 50% primary air. Therefore, the shutter should never be closed down completely since this closes off the source of primary air.

PILOT BURNERS

The main burners are ignited by a pilot burner, which is just a small gas burner. It is connected to the main gas valve with a piece of tubing which feeds a small amount of gas to the pilot burner. The pilot burner is slotted, allowing the flame to extend out several inches in two directions, Figure 5-9, and mounted so that one of the flames will be directed to the crossover igniter for the main burners. Thus, when gas is supplied to the main burners, all will light simultaneously. Primary air for the pilot is supplied by a slot or drilled hole in the burner head.

FIG. 5-10. Burner and thermocouple. *Courtesy ITT.*

The pilot safety or thermocouple, Figure 5-10, is connected to the pilot burner assembly. A thermocouple is an electric generator which determines whether or not the pilot light is lit. It would be dangerous to open the main gas valve without any pilot to ignite the gas as it enters the main burner. Gas could collect in the furnace and surrounding area and create a very definite explosion and safety hazard. To prevent this, the pilot flame is directed to the top of the thermocouple and the current generated in the thermocouple tells the gas valve that the pilot is lit. How this is accomplished will be covered in a later section.

PILOT MAINTENANCE

All furnaces show the procedure for lighting the pilot burner on the AGA nameplate attached to the furnace. To light the pilot, first set the room thermostat below the point where it will call for heat or pull the furnace disconnect switch so that the furnace cannot come on when the pilot is lighted. The manual valve on the gas valve should be turned to its *pilot* position and then depressed and held down, Figure 5-11. Light the pilot by holding a match to the gas outlet of the pilot, and hold down the manual valve about 60 seconds to allow the thermocouple to heat up. After this time period, the manual gas valve should be rotated to its *on* position and released.

Some gas valves will have a plunger, usually colored red, which must be held in to light the pilot. In either case, the procedure is essentially the same.

If seasonal or temporary shutoff is

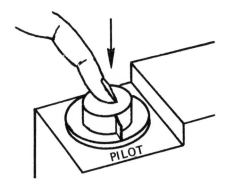

FIG. 5-11.   Depressing the manual valve before lighting pilot.

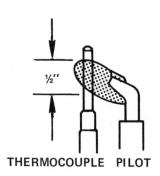

THERMOCOUPLE   PILOT

FIG. 5-12.   Pilot flame should surround 1/2 inch of the thermocouple.

desired, the knob can be turned back to *pilot*. If the knob is turned to the *off* position, there is a safety circuit on many valves which will not allow the knob to be turned back to *pilot* for 3-5 minutes. *Do not force the knob!*

The pilot flame should be observed. It should be a soft blue flame and should surround the top $\frac{3}{8}$ to $\frac{1}{2}$ in. of the thermocouple tips, Figure 5-12. If it does not, this

could be due to improper location of the pilot or thermocouple, a dirty orifice in the pilot, or low gas pressure. The gas pressure to the pilot flame can be adjusted by turning a screw on the gas valve or, if there is a separate pressure regulator, it can be adjusted at this point.

Baso Division of Penn Controls offers the following flame characteristics as evidence of trouble, Figure 5-13.

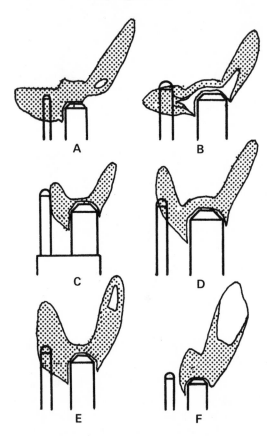

FIG. 5-13.   Flame shapes that indicate malfunctions.

A. Lazy Yellow Flame on the pilot burner, when the combustion chamber is cool and the gas pressure is normal, indicates a dirty pilot burner. When pilot burner is clean and combustion chamber is hot, the lazy yellow flame indicates high ambient temperatures.

B. Lazy Blue Flame, which has no well-defined shape and waves about, probably indicates burning at the orifice. This condition may also be betrayed by a high-pitched, whistling sound characteristic of burning at the orifice.

C. Very Small Flame, when valves are wide open and pressure is normal, shows that the orifice is clogged or that the pilot burner is filled with carbon produced by burning at the orifice. This carbonization is peculiar to manufactured gas but clogging of the orifice may occur with all gases. If the appliance uses a filter in the pilot line, the very small flame may be caused by a dirty filter or by one that is too small for the service demand put upon it. If the pilot flame becomes very small when the main burner comes on, it may indicate a clogged pilot filter, too small a manifold, or insufficient pressure due to some other cause. Enlarging the pilot orifice may sometimes give temporary relief.

D. Lifting, Blowing Flame is noisy and will sometimes blow itself out. This condition usually is caused by high pressure or by too large an orifice or by both. Correct by reducing the pressure to normal range and by using the proper type and size of orifice inlet fitting. A noisy, roaring flame may result from placing the pilot in a strong stream of inrushing air. Proper shielding of the pilot from this airstream impingement will correct the trouble.

E. Hard, Sharp Flames are peculiar to manufactured, butane-air, and propane-air gases. The flame will be noisy and have a tendency to blow itself out or to backfire and burn at the orifice. Correct by using a dual orifice inlet fitting of proper type and size. A temporary correction may be made by increasing the size of the orifice and/or decreasing the size of the primary air hole.

F. Yellow Flame Tips indicate too little primary air. The primary air opening may be too small or may be clogged with dirt or lint. See that the proper type and size of inlet fitting is installed. With LP gas, small yellow tips are not a matter of concern.

If burners become fouled with lint or dirt, they should be removed and cleaned. To remove the burner, first shut off the main gas supply and remove the screws holding the gas manifold in place. The pilot gas line and thermocouple leads to the gas valve should be removed at the gas valve, and the manifold rotated to remove the burners from the heat exchanger. Burner ports can be cleaned with a wire brush. If they are very dirty, a small piece of sheet metal, which will fit into the burner ports, can be used to clear each port. The inside of the burner can be cleaned with a stiff, bottle-type cleaning brush which is long enough to reach to the end. Burners can then be replaced in the heat exchanger, taking care to relocate the pilot and thermocouple in their original location. Crossover igniters should be checked to see if they line up properly and burner operation and ignition should be checked after reinstallation.

THERMOCOUPLES

In a gas furnace, the main gas valve must stay closed if the pilot is out for any reason or there is no gas supply to the pilot burner. Otherwise, gas will spill up the flue and into the furnace area, creating a dangerous, explosive atmosphere. The device used to prevent this is a thermocouple, Figure 5-14.

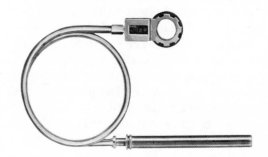

FIG. 5-14. Typical thermocouples. (Above) *Courtesy Honeywell, Inc.* (Below) *Courtesy ITT.*

A thermocouple is a bond of two dissimilar metals which, when heated, generate a small DC voltage. The voltage produced is measured in millivolts or 1/1000 of a volt (.001 volt). There are special millivoltmeters available for measuring DC voltages. Some regular volt-ohmmeters also have this capability, Figure 5-15. These meters have two leads, just as a volt-ohmmeter introduced in Basic Electricity. They are used and read in exactly the same manner.

A thermocouple has a tubular piece surrounding an inner solid element. The two are welded together at one end which is heated by the pilot burner flame and called the *hot junction*. The outer element is brazed to a brass connector sleeve which, in turn, is brazed to a copper tube. Inside the copper tube is an asbestos-insulated copper wire which is welded to the inner element of the thermocouple. The outer tube and inner wire form the *cold junction* of the thermocouple. When the *hot junction* is heated by the pilot, while the *cold junction* remains at a lower temperature, an electrical current is set up. The amount of current produced is, within limits, proportional to the difference in temperature between the hot and cold junctions. The current is fed to an electromagnet whose magnetic field opens the pilot gas valve and allows gas to flow. As long as heat is applied to the hot junction, current will flow and the pilot valve will remain open, Figure 5-16.

Since the *hot junction* must heat up before it will generate enough voltage to hold open the pilot gas valve, a plunger is provided to manually open the valve and

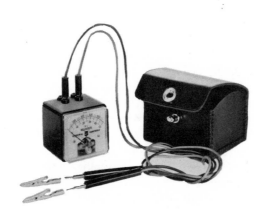

FIG. 5-15. Typical DC millivoltmeter. *Courtesy ITT.*

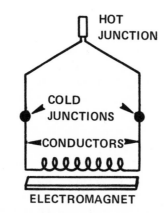

FIG. 5-16. Schematic showing thermocouple operation.

allow gas to flow to the pilot burner. Once the pilot is ignited, the plunger is held in manually for about 60 seconds to allow the generated voltage to build up to the point where the coil will hold the valve open.

If the pilot goes out for any reason, the generated voltage will drop to zero, the valve will close, and no gas will reach either the pilot or main burners because both the pilot and main valves are closed. Even if the main gas valve were to open on a call for heat from the thermostat, no gas would be available for the main burners because the pilot valve is closed. The pilot must be manually reset to restore service.

This system provides 100% safety shutoff, which is required with LP and an added safety factor with natural gas. It is most important with LP because LP is heavier than air and, if unburned gas were delivered to the furnace, it would drop to the floor and present a highly-dangerous explosion hazard. Natural gas, being lighter than air, would tend to rise and be somewhat diluted with room air.

Note that this safety circuit is *completely independent* of any outside power source and will function as long as the pilot stays lit, even with all power cut off from the furnace.

The thermocouple should be checked under closed circuit conditions. For this check, a General Controls Adapter No. 103050G, Figure 5-17, is used to provide

FIG. 5-17. General Controls Adapter allows the millivoltage to be measured under load. *Courtesy General Controls ITT.*

connection points for the millivoltmeter, allowing the circuit millivoltage to be read under load. The thermocouple should be removed from the gas valve and this adapter screwed into the hole, finger-tight. The thermocouple is then attached to the top of the adapter, finger-tight plus ¼ turn with a small wrench. The positive lead of the multi-tester is placed on the outside conductor of the thermocouple and the negative lead to the tab of the adapter, Figure 5-18. Use the 0-50 millivolt scale. If the meter moves to the left of zero or no reading is indicated, reverse the probes. All readings should be taken with the pilot burning.

If the reading is less than 7 millivolts, the following steps can be taken:

1. Adjust the pilot gas for a larger flame. This can be done at the gas valve by turning the pilot flame adjustment screw counterclockwise, Figure 5-19.
2. Clean the primary air holes in the pilot burner.
3. Clean the pilot burner orifice.

If the reading is still less than 7 millivolts, the thermocouple should be replaced.

GAS VALVES

The most commonly used gas valves will include a pressure regulator, pilot valve and main gas valve, Figure 5-20. They will have a tap for taking pressure readings and a choice of inlet and outlet connections. They will also have a screw to adjust manifold pressure and pilot flame size. The thermocouple attaches to the gas valve with a threaded connection and spade or screw terminals are provided for the main gas valve coil connections.

FIG. 5-18. Connecting millivoltmeter probes to adapter.

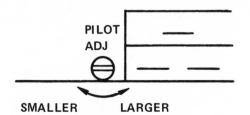

FIG. 5-19. Adjusting pilot gas supply.

To assure proper piping, the direction of flow through the gas valve will be indicated by an arrow on the valve body. Several outlet options are often available. For instance, the Honeywell V800 has a left. right or straight through outlet to the manifold. A good-quality pipe dope should be used on the male connection, leaving the last two threads bare. Pipe should not be inserted too far—just up to the last two imperfect threads.

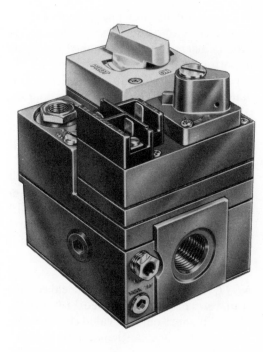

FIG. 5-20. Typical gas valves. (Above) *Courtesy Honeywell, Inc.* (Below) *Courtesy General Controls ITT.*

The pilot burner gas supply tube is connected with a compression fitting. The tube is inserted into the tapped hole until it bottoms. While holding tubing all the way in, slide the compression fitting into place and engage the threads, turning until finger-tight. Then, with a wrench, tighten one more complete turn. This breaks off the ferrule and clinches the tubing.

After installation, be sure to check all pipe joints, pilot gas tube connections and valve gaskets for leakage. Paint them with a rich soap and water solution and watch for bubbles while the main burner is in operation.

NOTE   In making the electrical connections, *never* jumper the coil terminals. This will short out the valve coil and may burn out the heat anticipator in the thermostat.

Some recent valves provide for two-stage heating and are used with a two-stage heating thermostat. The first stage activates a preset, low-fire position. The valve opens to the high-fire position when the second stage calls for heat. The shift is accomplished by a heat motor in the valve operator. This type of valve requires about 30 seconds to recycle.

The simplified block diagram will show the basic gas valve sequence of operation as follows:

1. Dial in *off* position (Figure 5-21). No gas can flow to either the pilot or main burner section.

2. Dial in *pilot* position (Figure 5-21). Gas can flow through the dial passageway

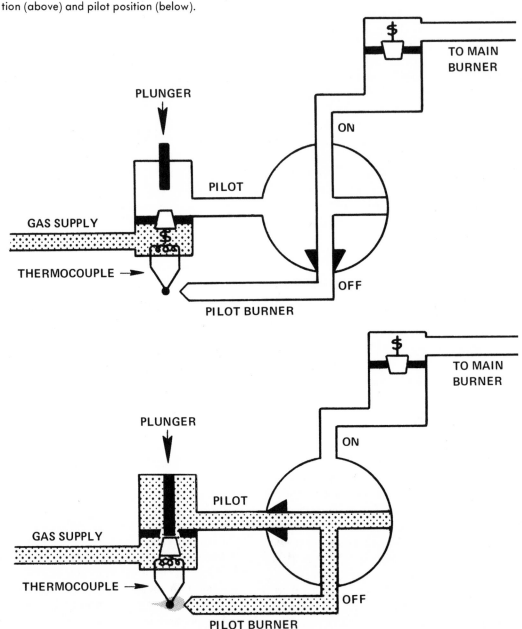

FIG. 5-21.   Schematic of gas valve in off position (above) and pilot position (below).

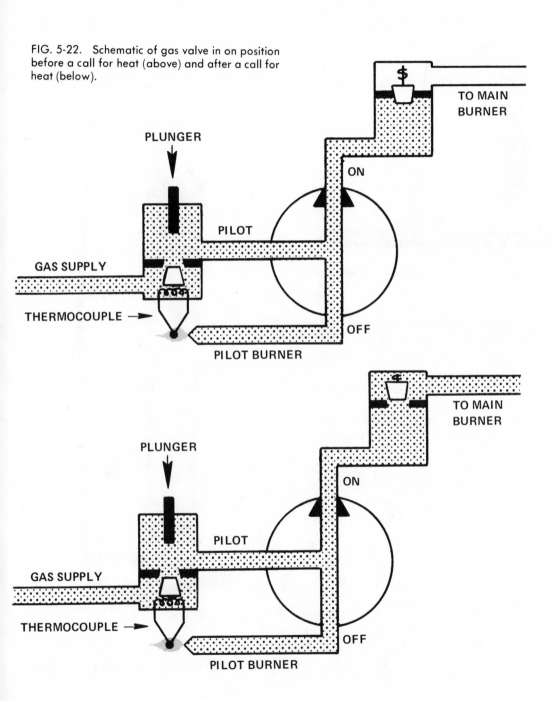

FIG. 5-22. Schematic of gas valve in on position before a call for heat (above) and after a call for heat (below).

to the pilot burner valve. Pushing the plunger opens the valve manually, allowing gas to flow to the pilot burner. Note that no gas can flow to the main gas valve section, so the main burners cannot come on, even after a call for heat opens the main gas valve. After about 60 seconds, the coil will be energized, holding the pilot gas valve open.

3. Dial in *on* position, (Figure 5-22). Gas can now flow through the dial passageway to the pilot burner valve (which is open when the pilot is lit) and to the main burner valve. On a call for heat from the thermostat, the main burner valve opens, supplying gas to the main burners for ignition by the pilot.

SAFETY DROPOUT

To check whether the safety dropout in the gas valve is functioning correctly, and not sticking, requires a stop watch or a wrist watch with a second hand. Begin timing when the pilot light is turned to the *off* position. A click will indicate when the gas valve has closed off the main gas supply. The maximum allowable time for the safety dropout to close is 2½ minutes. If it takes longer for the valve to drop out, it indicates a sticky valve which should be replaced.

A millivoltage meter should be attached to the thermocouple adapter during this check. The normal reading will be below 4 millivolts. If the millivoltage reading is not 4 millivolts or below when safety dropout occurs, the millivoltage coil in the main gas valve is becoming weak. In most cases, the valve must be replaced to correct this problem.

After making the check, the meter leads are removed and the thermocouple adapter removed from the gas valve and thermocouple. Replace the thermocouple in the gas valve, tightening it finger-tight plus ¼ turn with a light wrench. Relight the pilot and turn the gas valve to the *on* position.

MANIFOLD GAS PRESSURE

The pressure regulator is part of the combination gas valve. Earlier units installed a separate regulator in the line leading to the main gas valve. It is necessary to keep gas pressure to the burners constant, even with variations at the mains. At meal times, for instance, there is a drop in pressure at the mains because of increased usage.

The gas valve opens relatively fast but the regulator opens slowly. This insures quiet light-up and prevents temporary starvation of the pilot.

A simple manifold pressure test kit—containing a *B* valve or petcock which screws into the test port on the gas valve or manifold, a 4 ft length of rubber tubing and a pressure gauge—should be in every serviceman's tool kit.

Because manifold pressures are low (less than one pound per square inch) they are measured in inches *water gauge*, abbreviated *in. w.g.* Manifold pressure for natural gas should be between 3½ and 4" w.g. and for propane, 11" w.g.

The pressure in the manifold is checked in the following way: First turn off the furnace disconnect switch to remove all power from the gas valve so that it cannot open and allow gas to leak

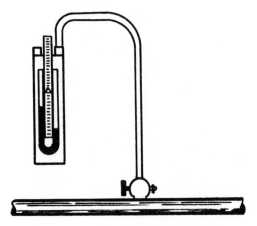

FIG. 5-23. U-tube manometer attached to a *B* valve.

into the room while the plug is out. There will be an ⅛" pipe plug on the gas valve or manifold and this should be removed. The *B* valve is installed into this plug tap with the valve in the *off* position, Figure 5-23. Pressure can be taken with either a U-tube manometer or manifold gas pressure gauge, the hose fitting being attached to the *B* valve.

The U-tube manometer, Figure 5-24, has graduated scale in the center of the U, and shutoff tubing connectors which open into the top of each leg. These are turned one turn counterclockwise to open. The manometer has a magnet on the assembly so that it will adhere to any smooth steel surface. It also can be hung vertically. The U-tube contains a special fluid and, with equal pressure on both tubes, the height of the liquid in both tubes is the same. The center scale is adjusted so that the zero mark is set at this point. The rubber or nylon tubing connects one of the shutoff

FIG. 5-24. Typical U-tube manometer. *Courtesy Bachrach Instrument Co., Div. of AMBAC Industries.*

openings to the *B* valve. When the valve is opened, pressure is applied to one column of the fluid. This will force that column down and the other up. The *difference* in pressures, that is the sum of the amount one column is below zero and the amount the other is above, is the pressure reading, Figure 5-25.

Note that some manometers will have the zero setting at the bottom of the scale and numbers on one side only. With these, the opening on the side which does *not* have the calibration is connected to the *B* valve. The pressure in inches of *water gauge* is read directly from the height of the left hand column.

With the thermostat calling for heat, turn on the disconnect switch and open the *B* valve. The pressure reading should be between 3 and 4" w.g. for natural and 11" w.g. for LP. If above or below this point, the pressure regulator should be adjusted. Gas valves with a built-in pressure regulator will have an adjustment screw on the gas valve to change the pressure. This screw can be turned up or down to increase or decrease pressure (clockwise increases pressure). A quarter turn or less should be sufficient to correct the pressure reading. If the unit has a separate pressure regulator, it will have an adjustment screw on the top which can be turned in the same manner.

The pressure regulator usually supplied for natural and other gases is not used for LP gas under normal circumstances. The reason is that there must be a master regulator at the LP supply tank to maintain a constant downstream pressure of 10 to 12" w.g. Since this regulator is

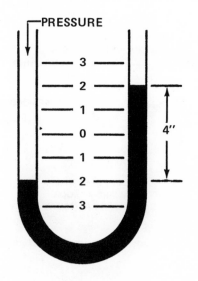

FIG. 5-25.   Reading a manometer.

always installed, it is not necessary to duplicate its function at the furnace.

After the reading is taken and the pressure regulated, the disconnect should be turned off, the gauge, *B* valve and U-tube removed, and the plug put back in the gas manifold. Apply sealing compound to the plug and double-check with soap bubbles to be sure that it does not leak.

CONTROL CIRCUIT AMP-DRAW

A method of checking the control circuit amp-draw at the thermostat, using an amperage multiplier, was described under **Thermostats. The same value can be taken** at the gas valve (on up-flo furnaces only) by removing one of the low voltage wires from the gas valve terminal. One end of the amperage multiplier is connected to this wire and the other end to the terminal from which the wire was disconnected. The furnace disconnect can be turned on and the reading recorded with an Amprobe, whose jaws are opened and placed around the coil of the amperage multiplier. The actual amperage is calculated by dividing this reading by 10. The disconnect switch is then turned off, the amperage multiplier removed and the low voltage wire reconnected to the gas valve. The disconnect switch is then turned on.

TEMPERATURE RISE

Temperature rise is the number of degrees the air is heated as it passes through the furnace. Most gas furnaces are designed to have a temperature rise from 80 to 100° F. Temperature rise is measured by inserting a thermometer in the return air side of the furnace and reading the temperature of the return air. This will normally be in the range of 68 to 72° F. A second reading is taken in the supply air plenum. Care should be taken to assure that the supply air reading is not influenced by radiant heat from the heat exchanger. To avoid this, the supply air reading should be taken at a point which is out of the line-of-sight of the heat exchanger, Figure 5-26. The supply air should be about 90° greater than the return air. For example: with a return air temperature of 70° F, the supply air at the plenum should be about 160° F.

The supply air temperature can also be taken at a supply air register. If taken there, an allowance of about ½° F per foot of duct, from the plenum, must be used to compensate for heat loss in the ducts.

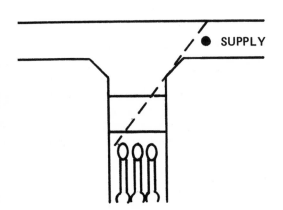

FIG. 5-26. Supply air temperature should be measured at a point out of the line-of-sight of the heat exchanger.

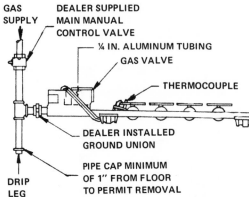

FIG. 5-27. Gas piping layout, including drip leg.

If the temperature rise is below 90°, then the blower should be slowed down to increase the heat rise through the furnace. If the rise is greater than 90°, then the blower should be speeded up to reduce the rise. (One full turn of the outer half of the adjustable pulley will change the temperature rise through the furnace about 10 to 15°.)

GAS PIPING

It is most important to have correctly sized piping to the gas unit. The relevant Code of Practice, *Gas Installation Pipes*, should be consulted to determine the size of piping to be used. The pipe should be run direct from the meter to the unit with as few bends as possible. The riser from the burner should consist of a tee fitting with the bottom outlet capped and extended to form a drip leg. This prevents dirt and dust from being carried into the burners, Figure 5-27. During installation, all piping should be cleaned of dirt and scale by hammering it on the outside. Ends of piping and tubing should be carefully reamed to remove obstructions and burrs. Nonhardening pipe joint compound should be used sparingly and on the male threads only. All joints should be carefully checked for leaks with a soap solution to be absolutely certain they are gas tight.

PIPE SIZES FOR GAS HEATERS

Most authorities follow the recommendations shown in Table 5-1 in specifying the proper size of gas piping to use for gas heaters. These recommendations are based on a system of piping that, as a whole, is designed for a pressure drop of 0.3" w.g. from the meter to the heaters. Table 5-1 shows the capacity, in cubic feet per hour, for various sizes of pipes in various lengths. These capacities are based on using gas with a specific gravity of 0.60. (If the gas being used has a differ-ent specific gravity, adjust the value from Table 5-1 by the multiplier in Table 5-2.) By adopting the rather low pressure drop of 0.3", there is sufficient allowance for the extra resistance of the ordinary number of fittings used in the piping. In other words, it is only necessary to measure the length of the gas piping from the meter to the heater, disregarding the extra resistance imposed by the normal number of fittings, to determine the length that will be used in the foregoing table to find the carrying capacity of pipes. In areas where high gas pressure is always available at the house service, a pressure drop greater than 0.3" w.g. is permissible. Natural gas heaters are usually designed for a 3½ to 4" w.g. manifold pressure. The pressure drop through the heater controls must never exceed 1.0". If 8.0" of pressure is always available at the meter, piping could be designed for as much as 1.0" pressure drop and still have adequate pressure available at the heater. In areas where such generous gas pressure is always available, the local regulations have usually allowed for this in their recommendations on pipe sizing.

LP GAS TANKS AND PIPING

The chapter on Combustion noted that the LP gases, butane and propane, have specific gravities of 1.53 and 2.01 respectively. Both are shipped and stored in liquid form under about 200 lbs of pressure. To be used in a furnace, they must be vaporized. While a combination gas valve for natural gas includes a pressure regulator to maintain gas pressure, this is not necessary for an LP installation, since the proper pressure is maintained before the gas enters the house. The LP gas pressure must be 11" w.g. at the unit, under full load, in order to provide satisfactory operation.

The piping for LP gas units can be sized, using Tables 5-1 and applying the multipliers, from Table 5-2, for the differ-

TABLE 5-1. Pipe Capacity Cubic Feet of Gas per Hour (0.60 Specific Gravity)

Length of Pipe in Ft	Nominal Diameter of Pipe (Inches)				
	½	¾	1	1¼	1½
15	76	172	345	750	1220
30	55	120	241	535	850
45	44	99	199	435	700
60	38	86	173	380	610
75		77	155	345	545
90		70	141	310	490
105		65	131	285	450

TABLE 5-2.
Multipliers for Table 5-1 When the Specific Gravity of Gas Is Other Than 0.60

Specific Gravity	Multiplier
.35	1.31
.40	1.23
.45	1.16
.50	1.10
.55	1.04
.60	1.00
.65	.962
.70	.926
.75	.895
.80	.867
.85	.841
.90	.817
1.50	.633
1.55	.622
2.00	.547

TABLE 5-3.

Use this size copper tubing or standard pipe to keep pressure drop below 2 lbs for the maximum flow shown if the line between regulators (tank to building) is this long:

Capacity	25 Feet	50 Feet	75 Feet	100 Feet
50 cfh 125,000 Btuh	⅜" OD Tubing	⅜" OD Tubing	⅜" OD Tubing	⅜" OD Tubing
100 cfh 250,000 Btuh	⅜" OD Tubing	⅜" OD Tubing	⅜" OD Tubing	½" OD Tubing
150 cfh 375,000 Btuh	½" OD Tubing	½" OD Tubing	½" OD Tubing	½" OD Tubing
200 cfh 500,000 Btuh	½" OD Tubing	½" OD Tubing	½" OD Tubing	½" OD Tubing
300 cfh 750,000 Btuh	½" OD Tubing	¾" Pipe	¾" Pipe	¾" Pipe

TABLE 5-4.

Use this size copper tubing or standard pipe to keep pressure down below ½" water column for the maximum flow shown if the line from the second stage regulator to the appliance is this long:

Capacity	10 Feet	20 Feet	30 Feet	40 Feet	50 Feet
10 cfh 25,000 Btuh	⅜" OD Tubing	⅜" OD Tubing	½" OD Tubing	½" OD Tubing	½" OD Tubing
20 cfh 50,000 Btuh	½" OD Tubing	½" OD Tubing	½" OD Tubing	⅝" OD Tubing	⅝" OD Tubing
30 cfh 75,000 Btuh	½" OD Tubing	⅝" OD Tubing	⅝" OD Tubing	⅝" OD Tubing	⅝" OD Tubing
50 cfh 125,000 Btuh	⅝" OD Tubing	⅝" OD Tubing	¾" Pipe	¾" Pipe	¾" Pipe
75 cfh 187,500 Btuh	¾" Pipe	¾" Pipe	¾" Pipe	¾" Pipe	¾" Pipe

TABLE 5-5.

Recommended tank size if your lowest outdoor temperature (average for 24 hour period) is:

Capacity	32°	20°	10°	0°	—10°	—20°	—30°
50 cfh 125,000 Btuh	115 gal	115 gal	115 gal	250 gal	250 gal	400 gal	600 gal
100 cfh 250,000 Btuh	250 gal	250 gal	250 gal	400 gal	500 gal	1000 gal	1500 gal
150 cfh 375,000 Btuh	300 gal	400 gal	500 gal	500 gal	1000 gal	1500 gal	2500 gal
200 cfh 500,000 Btuh	400 gal	500 gal	750 gal	1000 gal	1200 gal	2000 gal	3500 gal
300 cfh 750,000 Btuh	750 gal	1000 gal	1500 gal	2000 gal	2500 gal	4000 gal	5000 gal

ent specific gravity of LP gas. Another method of calculating the piping sizes is shown in Tables 5-3 and 5-4. LP gas equipment installation is covered under the safety standards of the National Fire Protection Association as published in their Manual 58. Maintaining the proper gas pressure depends upon three main factors:

1. Vaporization rate. The vaporization rate is determined by the outside air temperature and the wetted surface of the tank. Wetted surface is defined as the surface area of the tank which is covered by liquid. LP gas is vaporized by heat from the outside air passing through the walls of the tank where it comes in contact with the liquid fuel in the tank. The greater the wetted area, the greater will be the heat transfer. Therefore, high liquid levels are required for adequate vaporization. If the tank is nearly empty, very little heat

transfer and inadequate vaporization will occur.

The size of the tank becomes a factor in cold weather when air temperatures around the tank are low. If the tank is too small, there will not be enough wall area to collect the heat required to vaporize all the fuel consumed at extreme temperatures. Table 5-5 shows the sizes of tanks required to vaporize the gas consumed at various winter temperatures. These ratings assume that the tank is at least half full. Notice, in Table 5-5, that a 250 gallon tank vaporizes 100 cubic feet of gas per hour at 10° F and that this drops off to 50 cubic feet per hour at -10° F.

2. Proper pressure regulation. Many LP gas installations use the single-stage method where one regulator, at the tank, reduces tank pressure from about 200 lbs to the 11" w.g. required at the furnace,

Figure 5-28. This creates problems because a small variation in the pressure drop from tank to house can be a frequent source of customer complaints and service calls. A few inches of pressure drop below 11" w.g. upsets the normal operation of the furnace and a single-stage system can easily vary this much. In addition, because of the small outlet required, moisture in the fuel can freeze in the regulator.

A better method is to use two-stage pressure regulation, Figure 5-29. The first regulator, at the tank, reduces the 200 lbs pressure in the tank to between 5 and 15 lbs pressure in the intermediate line.

Manual 58 of the NFPA says that the maximum pressure which can be taken into a building is 20 lbs. Therefore, the line from tank to house should carry no more than 20 lbs pressure. An intermediate line pressure of about 10 lbs will provide sufficient capacity. The first stage regulator must have strong parts, a strong diaphragm and a large orifice built for handling pounds-to-pounds pressure. It must be capable of receiving up to 250 lbs inlet pressure and the outlet pressure should be incapable of being adjusted to over 20 lbs.

A second-stage regulator should have a 3/16 or 1/4 in. orifice with an inlet pressure of approximately 10 lbs and an outlet pressure of the 11" w.g. desired within the house. A variance of 3 or 4 lbs between the first and second stage is not critical and it is best to locate the second-stage regulator outside of the house. It must be 5 ft away from the closest lower opening to the house and the vent should be turned downward, Figure 5-30.

The two-stage system will give more

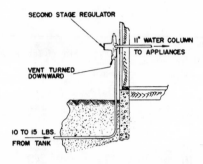

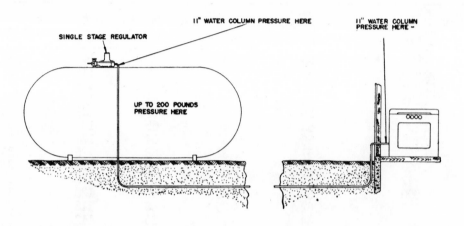

FIG. 5-28.   LP gas installation with single-stage regulator. *Courtesy Fisher Controls Co.*

FIG. 5-30.   Installing the second-stage regulator. *Courtesy Fisher Controls Co.*

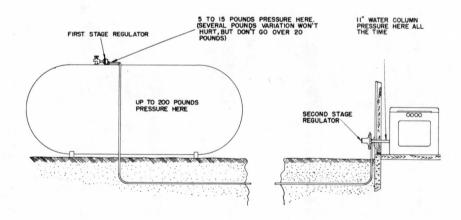

FIG. 5-29.   LP gas installation with two-stage regulator. *Courtesy Fisher Controls Co.*

uniform operating pressure, will reduce the possibility of freezing and will also reduce the size of the piping from the tank to the house. Therefore it has many advantages over single-stage systems.

3. Pressure drop in lines. The last factor for proper gas pressure is the pressure drop in the lines between the regulators and between the second-stage regulator and the furnace.

PROPANE PIPE DOPE

It should be remembered that propane is an excellent solvent. Therefore, when assembling a propane piping system, special pipe dope must be used. LP will dissolve white lead or most standard commercial compounds, so shellac-based compounds such as Gasolac, Stalactic, Clydes or John Crane are recommended for use. In many areas, gas companies will mix LP gas with other gases during peak periods so it is good practice to use these compounds for any gas piping.

TABLE 5-6. Gas Input to Burner in Cubic Feet per Hour

Seconds For One Revolution	Size of Test Meter Dial (Cubic Feet)				Seconds For One Revolution	Size of Test Meter Dial (Cubic Feet)			
	One-half	One	Two	Five		One-half	One	Two	Five
10	180	360	720	1,800	50	36	72	144	360
11	164	327	655	1,636	51	35	71	141	353
12	150	300	600	1,500	52	35	69	138	346
13	138	277	555	1,385	53	34	68	136	340
14	129	257	514	1,286	54	33	67	133	333
15	120	240	480	1,200	55	33	65	131	327
16	112	225	450	1,125	56	32	64	129	321
17	106	212	424	1,059	57	32	63	126	316
18	100	200	400	1,000	58	31	62	124	310
19	95	189	379	947	59	30	61	122	305
20	90	180	360	900	60	30	60	120	300
21	86	171	343	857	62	29	58	116	290
22	82	164	327	818	64	29	56	112	281
23	78	157	313	783	66	29	54	109	273
24	75	150	300	750	68	28	53	106	265
25	72	144	288	720	70	26	51	103	257
26	69	138	277	692	72	25	50	100	250
27	67	133	267	667	74	24	48	97	243
28	64	129	257	643	76	24	47	95	237
29	62	124	248	621	78	23	46	92	231
30	60	120	240	600	80	22	45	90	225
31	58	116	232	581	82	22	44	88	220
32	56	113	225	563	84	21	43	86	214
33	55	109	218	545	86	21	42	84	209
34	53	106	212	529	88	20	41	82	205
35	51	103	206	514	90	20	40	80	200
36	50	100	200	500	94	19	38	76	192
37	49	97	195	486	98	18	37	74	184
38	47	95	189	474	100	18	36	72	180
39	46	92	185	462	104	17	35	69	173
40	45	90	180	450	108	17	35	67	167
41	44	88	176	440	112	16	32	64	161
42	43	86	172	430	116	15	31	62	155
43	42	84	167	420	120	15	30	60	150
44	41	82	164	410	130	14	28	55	138
45	40	80	160	400	140	13	26	51	129
46	39	78	157	391	150	12	24	48	120
47	38	77	153	383	160	11	22	45	112
48	37	75	150	375	170	11	21	42	106
49	37	73	147	367	180	10	20	40	100

To convert to Btu per hour, multiply by the Btu heating value of the gas used.

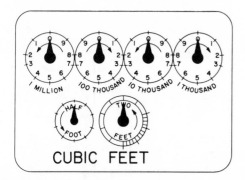

FIG. 5-31. Gas meter dials.

BURNER INPUT

To measure the input to a gas heater, the flow of gas at the gas meter can be timed and corrected to Btu input per hour. To do this accurately requires that all other gas appliances in the house be turned off. There is one hand on a gas meter, Figure 5-31, that measures small quantities of flow, like 1, 2 or 5 cu ft. The marking on the dial tells how many cubic feet of gas will pass through the meter for one complete revolution of the hand on that dial.

With the heater turned on, determine how much time is required for one revolution on the smallest dial. Table 5-6 will then show the cubic feet per hour the furnace will use. This is multiplied by the calorific value of the gas to obtain the Btu input. Heater output is usually taken as 80% of the input.

The average Btu per cubic foot of the gas, in the area of the installation, may be obtained from the local Gas Company.

For example: If the smallest dial measures 1 cu ft and it takes 33 seconds for it to make one revolution with only the furnace on then, from Table 5-6, the furnace will consume 109 cu ft of gas per hour. If the heat content of this gas is 1000 Btu/cu ft then the furnace *input* is 109,000 Btu. The *output* will be 87,200 Btu (.80 x 109,000).

FIELD WIRING

The first step in field wiring is to run a 120v, 2-wire service from the main breaker panel in the house to the furnace. The main circuit breaker or fuse should be sized in accordance with the instructions that come with the unit. Wire size is specified in the instructions and will be in accordance with the National Electric Code. Note that some areas require conduit for all or part of this service. Local codes should be checked, A disconnect switch (this can be a fused handy box with off-on switch, Figure 5-32), must be provided on or near the furnace. This is fused according to manufacturer's instructions. The two wires are connected to the line side of the box with the fuse in the hot leg (black wire). Two wires are run from the disconnect to the line voltage make-up box in the furnace. The black wire is connected to the fan control and transformer leads and the white wire is connected to the white leads from the blower and transformer, Figure 5-33.

Two low voltage wires (three if there is fan control at the thermostat) are run from the thermostat to the low voltage make-up box or terminal block in the furnace. The terminal block will be marked in the same manner as the thermostat. Thus R, W, and G (if used) from the thermostat will connect to the R, W and G terminals of the terminal block. This will complete the field wiring for a heating-only gas furnace.

FIG. 5-32. Fused handy box with off-on switch.

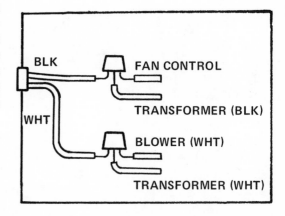

FIG. 5-33. Wiring connections in the furnace make-up box.

Questions

1. What is the purpose of the heat exchanger?
2. What is a simple check for heat exchanger leakage?
3. Name three considerations in good burner design.
4. How does the burner get primary air?
5. How does the burner get secondary air?
6. How is the amount of gas introduced in the burner controlled?
7. What is the burner venturi?
8. What is the purpose of the venturi?
9. How are several burners lighted simultaneously?
10. How are the main burners ignited?
11. What is a thermocouple?
12. Why is the thermocouple in the circuit?
13. Describe a properly operating flame.
14. Name six flame characteristics which are evidence of trouble. Describe the problems.
15. What is the principle operation of a thermocouple?
16. What is a millivoltmeter?
17. Describe the operation of a thermocouple.
18. What happens to the safety circuit if all power is cut off from the furnace?
19. What steps can be taken to correct a low millivoltage situation?
20. List the components of a combination gas valve.
21. How do you check for good piping installation?
22. What is safety dropout?
23. What is the maximum time allowable for the safety dropout to close and what would be the normal millivoltage reading at this point?
24. What is the normal manifold pressure for natural gas? Propane?
25. How is manifold pressure checked?
26. What is a U-tube manometer?
27. What is the normal temperature rise for a gas furnace?
28. How is temperature rise checked?
29. Name the three factors that affect gas pressure for an LP system.
30. In an LP system, why is the size of the tank important?
31. What are the advantages of a two-stage regulator system for LP?
32. How can burner input be measured?

Chapter 6
OIL HEATING

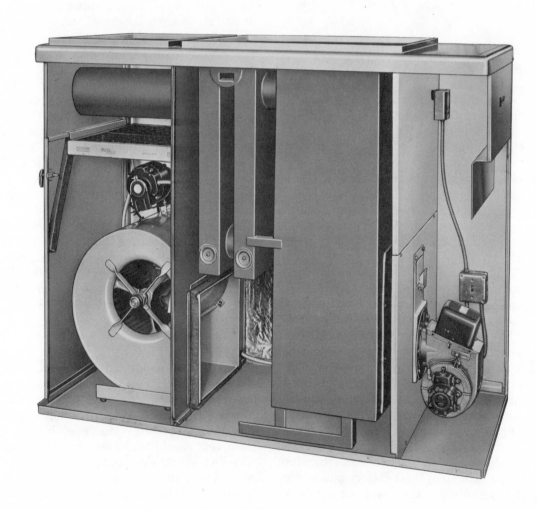

FIG. 6-1. Typical oil furnaces. *Exterior view courtesy International Environmental Corp. interior view courtesy Bard Mfg. Co.*

Many of the components common to all types of furnaces, regardless of fuel, were discussed in the previous section on Heating Components. These include the cabinets, filters, blowers, blower motors, **fan controls, limits and flues. An oil furnace has some special requirements, due to the fuel, which modify the design of the heat exchanger and the oil burner. Some typical oil furnaces are shown in Figure 6-1.**

The chapter on Heating Fundamentals demonstrated that oil in an open dish will not burn, due to the lack of sufficient air for ignition and complete combustion. Thus, an oil burner must put the oil in suspension and mix it with the proper amount of air for combustion. Further, it must maintain a fairly high temperature around the flame at all times. Impingement, or the touching of a flame on a cold surface, will result in incomplete combustion and form soot and other undesirable products. Oil furnace design, therefore, must take all of these factors into account and provide a means of overcoming them.

OIL

Most domestic oil burners will use No. 2 grade fuel oil because it is less expensive, has better lubricating properties and has a high Btu content (144,000 Btu/gallon). If the oil tank is outside, and the outside temperatures are extremely low for prolonged periods of time, then No. 1 oil is usually recommended. Gasoline, crankcase oil or any oil containing gasoline is positively not recommended for a domestic oil burner.

OIL STORAGE TANKS

Natural gas, of course, is supplied to the house by the utility, at a constant pressure through its own lines, and therefore no storage facilities are required. In contrast, fuel oil is delivered to the home in bulk and must be stored in sufficient quantity to last for a reasonable period of time, even in very severe weather. Fuel oil is commonly stored in a tank placed outside the home, above or below ground, or inside the basement, if this space is available. In placing and piping the tank, the local rules and regulations must be adhered to and, in the U.S., Underwriters' Laboratories Tank Specifications are the ones that apply.

GENERAL RULES

If the tank is buried outside, all the pipe connections must be made with swing joints so that, in case the tank settles, there will be no breakage. The top of the tank should be below the frost line. The tank vent should not be less than 1¼" pipe size and should slant toward the tank. It should terminate outside the building, at a point not less than 2 feet, vertically or horizontally, from any window or other building opening. The vent should have a weatherproof hood and should rise above the normal accumulation of snow and ice. Inside tanks should not be located within 7 feet, horizontally, of any fire or flame.

All pipe work and fittings must be airtight and only high grade materials should be used. Pipe lines should be run as directly as possible, be free of traps, and out of the way. If possible, they should be placed beneath the floor. Copper tubing is the most desirable and, on gravity flow, inside tanks, ⅜-inch OD is the most common for relatively short runs. Longer runs require ½-inch OD. On outside, underground tanks, where there is a suction lift, ½-inch OD tubing is recommended. On all installations requiring suction lift, the return line must be run at the proper angle in order to handle entrained air, in the oil or in the system, and return it to the tank. This must be done to assure proper fuel pump operation. The fill pipe should not be less than 2 inches and should slant toward the tank. It should terminate outside the building at a point not less than 4 feet from any building opening at the same or a lower level. Fuel terminals should be closed tight, provided with a metal cover and designed to prevent tampering. The tank will also have a shutoff valve at the exit and a filter between the shutoff valve and the furnace.

The tank may have an oil level indicator or it may have a plug so that the oil level can be checked with a dip stick. A standard size, vertical tank will hold 275 gallons of fuel oil. Figure 6-2 shows the approximate number of gallons in the tank when measured by inserting a dip

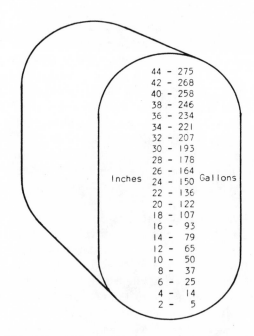

FIG. 6-2. Oil level table equates gallons to inches on dipstick.

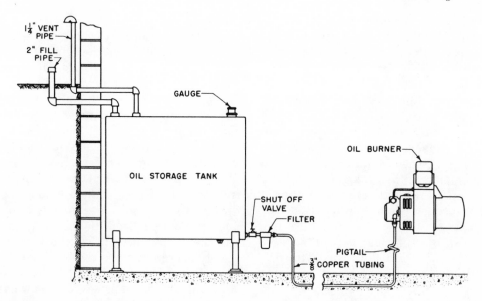

FIG. 6-3. Single pipe, gravity feed system is used where oil tank is above burner.

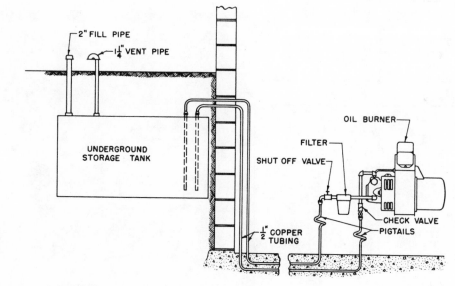

FIG. 6-4. Two pipe system must be used where vertical lift exceeds eight feet.

stick from the top. Water in the tank can be checked by putting some gray litmus paste on the bottom of the dip stick. This litmus paste will turn purple in the presence of water. If the water level is over 1 inch, it should be drained out of the tank.

A single pipe, gravity system is used where the oil tank is above the burner or where the lift does not exceed 8 feet. In a single pipe system, one line of ⅜-inch copper tubing is run from the oil tank shutoff valve to the oil burner, Figure 6-3. Excess oil is circulated internally in the pump, back to the suction side. In a gravity feed, the allowable length of ⅜-inch OD tubing is 100 feet. For a 5-foot lift, the allowable length is 90 feet and, with an 8-foot lift, this drops to 70 feet. A two pipe system can be used on any system but it *must* be used where the vertical lift exceeds 8 feet, Figure 6-4.

The return line is run back to the tank and, if the bottom of the tank is lower than the pump intake, the return line should be inserted in the tank so it is about 3 or 4 inches from the tank bottom. If the bottom of the tank is higher than the pump intake, the return line should extend not more than 8 inches inside the tank.

Allowable line lengths for various amounts of lift are shown in Table 6-1.

TABLE 6-1.   Recommended Line Lengths with a 3450 rpm Pump and ⅜ in. OD Tubing

One-line Lift / One-stage Pump	
Lift (ft)	Length (ft)
5	90
8	70

Two-line Lift / Two-Stage Pump	
Lift (ft)	Length (ft)
0	65
1	60
2	54
3	50
4	45
5	40
6	35
7	30
8	25
9	20
10	16

FILTERS

All installations will have an oil line filter located between the shutoff valve from the tank and the furnace. This filter will have a cartridge which can and should be replaced periodically to make sure that dirt or any other foreign materials will not enter the furnace burner, Figure 6-5. Replacement of this cartridge is quite simple and can be accomplished by first turning off the hand valve, shutting off the oil from the tank. Unscrew the bolt on the top of the filter assembly, remove the bowl and the cartridge from the bowl. A small pan should be placed below the filter assembly to prevent any leakage of oil on the floor. The inside of the bowl can be wiped clean and the gasket removed. A new gasket is supplied with the replacement cartridge. The cartridge is placed into the

FIG. 6-5.   Oil filter has a replaceable cartridge. *Courtesy Sundstrand Corp.*

bowl, the gasket is set on the lip and the mounting bolt firmly tightened. Be sure to open the hand valve after replacement is made.

Most of these filters have a bleed port so that any air entering the cartridge during the change can be bled off and not go through the entire system. The bleed port is opened slowly and at the same time that the oil valve is opened. This will bleed the air out of the filter. As soon as oil appears at the bleed hole, the port is closed.

OIL HEAT EXCHANGERS

Oil heat exchangers have a primary heating surface, or combustion chamber, plus a secondary heating surface. The reason for this is that the combustion chamber must come up to temperature rather quickly to support combustion. The secondary area provides enough metal surface for the air being delivered to the conditioned space to wipe across and absorb the heat from the flame.

Basically, combustion chambers will be round, square or rectangular. Those that are round can be either in the vertical or horizontal position. Round chambers are possibly the most popular design since they produce the least amount of air turbulence within the chamber itself. Several shapes of oil heat exchangers are shown in Figure 6-6.

A combustion chamber will be lined with either a light firebrick or a special wraparound, blanket-type insulating material which was originally designed for jet engines. There are several trade names for this, including Serafelt and Fiberfrax. This material is lightweight,

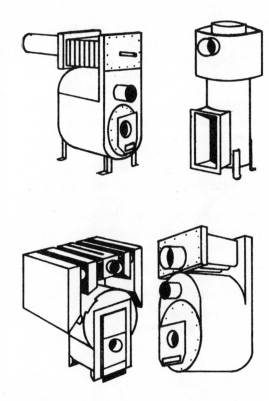

FIG. 6-6. *Typical oil heat exchangers.*

quite thin and can withstand temperatures up to 2500° F, which is well above the operating range of most domestic oil furnaces. The material comes up to temperature very quickly, in approximately 10 seconds, and it absorbs very little heat. The use of a refractory in the combustion chamber assures that the temperature in the chamber will remain high and thus aids in complete combustion of the oil and avoids impingement on a cold surface, Figure 6-7.

FIG. 6-7. *Refactory used in the combustion chamber. Courtesy Rheem Mfg. Co.*

The secondary heating surface can be either round or square. It will contain a series of baffles and channels so that the heated air and products of combustion will follow a labyrinth path, heating the entire surface evenly. It is also designed so that the supply air will come in contact with the entire outer surface for efficient heat exchange.

Most heat exchangers will have access doors so that the heat exchanger itself can be cleaned of any soot or carbon residue. Both the primary and secondary heat exchangers will have cleanout doors.

Since the size, shape and direction of

the flame is most important to the overall combustion efficiency of the furnace, an inspection hole is provided so the flame can be observed without opening the door. This peephole has a cover which can be slid aside to observe the flame. The door also is hinged so that, by using an inspection mirror, it is possible to observe the shape of a flame as it leaves the burner nozzle.

Heat exchangers will be made from cold-rolled or aluminized steel and, in many cases, they are coated with a ceramic to provide longer life in corrosive atmospheres.

FAN AND LIMIT CONTROLS

Oil furnaces, just like gas furnaces, will include either individual fan and limit controls or a combination fan and limit. Review the material on fan and limit controls, including secondary fan and limits, under Gas Heating. All of this material also applies to oil.

Fan controls are set to come *on* at about 100° F. With a differential of about 25° F, the *off* point is 75° F. The limit is set for 200° F. One difference between oil and gas is that frequently *both* the fan and limit are in the line voltage circuit of an oil furnace.

PRIMARY CONTROL

The safety control and flame detection circuits for an oil burner are both enclosed in what is called a *primary control*. This primary control has its own transformer and carries both line and low voltage. Two examples of primary controls are shown, Figure 6-8. Both primarily function in

FIG. 6-8. Two types of primary controls. (Above) *Courtesy Honeywell, Inc.* (Below) *Robertshaw Controls Co., Simicon Div.*

essentially the same manner with slightly different readings. Some are solid state and others include a solid state safety device and a regular relay for blower control.

The line voltage side of the primary control is connected to the ignition transformer and burner motor (orange wire). The hot or black wire is connected to the limit control (or both limits in the case of a down-flo furnace) and the white is connected to neutral. On the low voltage side, there will be 2 terminals marked T and T or T_1 and T for the thermostat connections (R and W from the thermostat) and two terminals for the flame detection device (Cad Cell) marked S and S, or FD and FD. The low voltage side does not have polarity and therefore, in both cases, either wire can be connected to either terminal. Figure 6-9 is a simplified wiring scheme as suggested by White-Rodgers.

With normal operation, limit controls would be in the closed position. On a call for heat from the thermostat, the ignition transformer and burner motor will be energized through the motor relay contacts, starting the motor and the spark ignition. If ignition is accomplished, the flame detector will take the heater, operating the safety circuit, out of the circuit and the burner will run until the thermostat is satisfied.

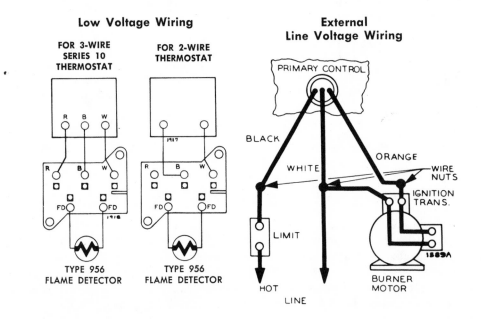

FIG. 6-9. Wiring diagram for the primary control. *Courtesy White-Rodgers Div., Emerson Electric Co.*

Here are some basic checks on a new installation or on a routine call:

1. Open the disconnect switch at the furnace.
2. Move the thermostat above the room setting so it will be calling for heat.
3. Make sure that there is a supply of oil to the burner.
4. Push the manual reset button (painted red) on top of the control.
5. Close the disconnect switch and let the burner run for about 5 minutes.
6. Remove one of the flame detector leads from the FD or S terminal. This control should lock out on safety, stopping the burner in approximately 15 seconds. This time will vary somewhat according to manufacturer's controls and should be checked.
7. Open the disconnect switch.
8. Replace the flame detector lead.
9. Turn on the power (furnace should not start).
10. Wait about 3 minutes and then push the manual reset button to bring on the burner.
11. The furnace should be cycled several times from the thermostat to be sure that it is operating properly.

These checks are all that are necessary if the control performs correctly and the burner continues to run on a call for heating. If the burner starts but then locks out on safety, the primary control and flame detector can be checked as follows:

1. Open the disconnect switch on the line voltage circuit to the furnace.
2. Remove the thermostat wires from the primary control terminals T and T.
3. Place a jumper wire across these terminals to complete the thermostat circuit.
4. Disconnect the two flame detector wires from the primary control.
5. Connect one end of a resistor to one of the flame detector terminals on the primary control. For White-Rodgers and Simicon, this resistor should be 2,000 ohms (red, black, red). For Honeywell, it should be 1,500 ohms (brown, green, red). Wattage is not important. Make sure that the other end is shaped in such a way that it can reach the other terminal easily and quickly, Figure 6-10.
6. Close the disconnect switch to the furnace and operate the red manual lever or button to start the burner.
7. Connect the other end of the resistor to the remaining flame detector terminal on the primary control. This must be done quickly to prevent the primary from going off on safety.
8. If the primary locks out on safety with the resistor in place, the primary can be considered defective. Be sure, however, that the resistor was connected quickly enough. If any doubt exists, wait 5 to 10 minutes and repeat the test. If the primary does not lock out with the resistor in place, the primary can be considered to be functioning properly.

FLAME DETECTOR

The oil burner control system must have a way of knowing whether or not ignition has been accomplished and a flame established and maintained during the heating cycle. If a flame has not been established, it is necessary to stop the supply of oil and turn off the furnace. This is accomplished by a flame detector which utilizes a Cadmium Sulfide Cell, or Cad Cell as it is normally called, Figure 6-11.

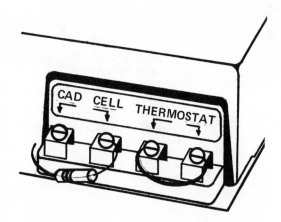

FIG. 6-10. Resistor wired to Cad Cell terminals.

FIG. 6-11. Flame detector uses a Cad Cell. *Courtesy Honeywell Inc.*

Cadmium sulfide has the unique capability of being a conductor of electricity in the presence of light in the visible range and will resist the passage of electricity in darkness. This capability is utilized to determine whether or not a flame is present.

The resistance of a Cad Cell to the flow of electrical current in darkness is about 100,000 ohms — great enough to prevent any flow. However, in visible light the resistance drops to less than 1,600 ohms which allows enough current to flow to pull in a sensitive relay in the primary control. This tells the primary that ignition has occurred and oil can continue to flow to the burner. A well-adjusted burner will cause the cell to operate in the range of 300 to 1000 ohms.

The Cadmium Sulfide Cell is located at the rear end of the burner tube in direct alignment with the flame from the nozzle, Figure 6-12. If the burner ignites, the Cad Cell will *see* this flame and the light will allow the Cad Cell to conduct electricity. The electrical circuit established then will drop out the heater in the safety circuit in the primary control, allowing the burner to run until the thermostat is satisfied. Placement of the Cad Cell, therefore, is very critical. It must see the light from the flame but not be *fooled* by transient light from other sources such as an open inspection door. The cell has been carefully positioned by the manufacturer and its location should not be disturbed. The considerations for selection of location are:

1. The cell must view the flame directly.
2. Light reaching the cell must produce ohm readings of less than 1,600, as described above.
3. The cell must not see either direct or reflected external light.
4. The ambient temperature must be below 140° F.
5. The cell must have adequate clearance so that metal surfaces near the cell will not affect it by movement, shielding or radiation.

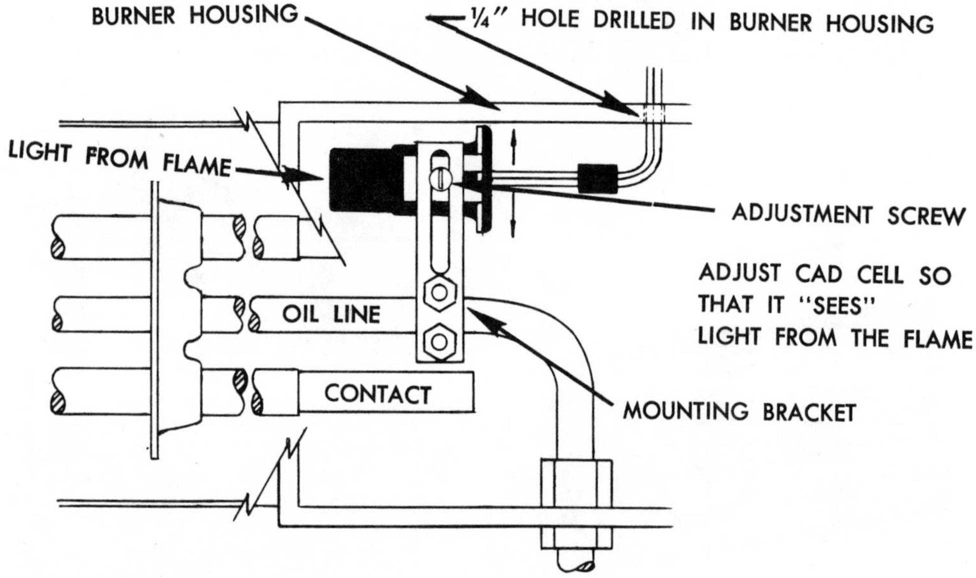

CHECKING THE FLAME DETECTOR

The flame detector can be checked with an ohmmeter set on the Rx10k scale. With the burner off, the flame detector leads are removed from the primary control and the resistance across these leads checked. This resistance should read over 100,000 ohms when the furnace is off.

To check the Cad Cell under normal operation, proceed as follows:

1. Jumper the thermostat terminals T and T on the primary control.
2. Push the red reset button and close the disconnect switch to start the burner.
3. When the burner comes on and is operative, disconnect the Cad Cell leads at the primary control. This should stop the burner in approximately 45 to 50 seconds. If the burner does not stop, then the primary control is defective and should be replaced.
4. Wait 2 or 3 minutes and then connect an ohmmeter to the two Cad Cell leads. Resistance should be 100,000 ohms or better. If not, the cell is

FIG. 6-12. Mounting the Cad Cell. *Courtesy Robertshaw Controls Co., Simicon Div.*

defective or there is light leakage into the burner and all access openings should be checked and sealed against light.

5. Turn off the disconnect switch and insert a jumper wire into one of the Cad Cell connections on the primary control. However, if both connections are jumpered at this point, the burner will not start.

6. Close the disconnect switch. The burner should not start until the reset lever on the primary control has been set.

7. Once the burner has started, complete the jumper connection to the other Cad Cell connection on the primary control.

8. With the burner running, connect an ohmmeter to the two leads of the Cad Cell and read the resistance. This resistance should not be greater than 1,500 ohms and ideally should be in the neighborhood of 600 or 700 ohms.

9. If the resistance is greater than 1,000 ohms, it indicates that the Cad Cell may be dirty, may be cracked or broken, may be misaligned in the burner, or it may have loose or broken wires.

10. Remove the Cad Cell and wipe the face with a soft rag and replace. Manufacturers attach the cell in various ways. Be careful in removing. If the situation still persists, then the flame detector is faulty and should be replaced. *Be sure to remove the jumper on the Cad Cell and be careful not to touch the thermostat wire to the flame detector terminal, since this can damage the primary control.*

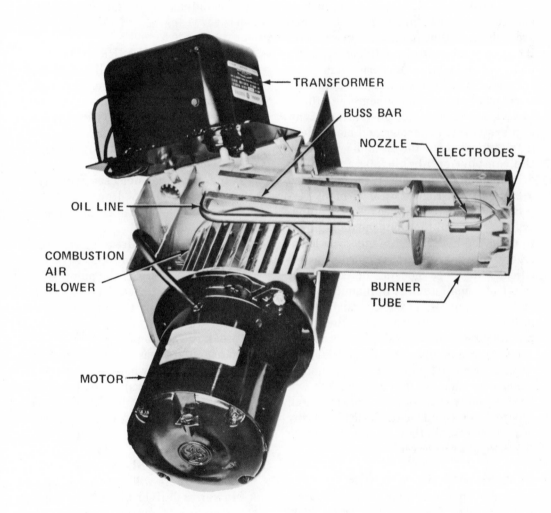

FIG. 6-13.   Cutaway view of an oil burner.
Courtesy Wayne Home Equipment Co., Inc.

GUN-TYPE BURNERS

As mentioned earlier, oil must be broken up or atomized in order for it to burn properly, and that it must be mixed with the proper amount of air to achieve complete combustion. A modern, high pressure, gun-type burner combines a number of components into one assembly to fulfill these requirements. Included are the ignition system (transformer and electrodes), an oil pump to put the oil under pressure, the combustion air blower, a motor to drive the pump and blower, the nozzle and a burner tube, Figure 6-13. In many respects, the industry has been able to standardize almost all the components required to build the high pressure gun-type burner, and most parts are assembled on a specifically designed housing or casting. A relatively small number of manufacturers make the pumps, nozzles, transformers, motors and so forth, and therefore maintenance and installation problems are very much simplified. Further, if a service mechanic can thoroughly understand one manufacturer's design, he will be reasonably successful in working on all types of this component. Basically, if any of these parts fail, they are not repaired in the field, but rather replaced.

BURNER ASSEMBLY MOTOR

The burner assembly motor can be either a 4 pole (1750 rpm) or a 2 pole (3500 rpm) motor. Its shaft connects directly to both the combustion air blower and the oil pump by means of a connector which is part of the assembly. It has a thermal overload and, if it goes off on the overload, it must be reset manually before it will operate. This motor should require only minor maintenance, including a few drops of oil every year.

CHECKING THE BURNER MOTOR

If the burner motor will not operate, reset the red safety button on this motor. If pushing this button starts the motor, then check for causes which may have shut it off on thermal overload.

1. Make sure that the supply voltage is between 110 and 120v.
2. If the motor has not been recently lubricated, use a few drops of SAE No. 20 nondetergent oil in each lubricating hole.
3. See if the pump is dragging the motor. Disengage the pump and rotate the shaft of the pump. If the pump is hard to turn, then it should be replaced in accordance with the pump instructions.
4. Check the combustion blower to see that it rotates easily and is not bound up or caught anywhere.
5. Rotate the motor shaft by hand. If it turns easily, but still will not operate electrically, the motor should be replaced.

COMBUSTION AIR BLOWER

All oil burners will have a combustion air blower to deliver air to the flame to assure complete combustion of the oil. Approximately 1,540 cubic feet of air is required to burn 1 gallon of oil. But, because it is not possible for the burner to mix all of the air with the oil, about 20 to 30% excess air is often required. Thus, most burners will furnish between 2,000 and 2,200 cubic feet of air per gallon of oil and some may run higher.

Basically, the amount of air delivered by the combustion air blower will determine the capacity of the burner. It is the amount of air delivered to the nozzle tip, and mixing with the oil, that determines how much oil can be burned. As the air is delivered to the point of the nozzle, where it will mix with the oil, it is desirable to have as much turbulence as possible. This will reduce the excess air needed for the mixing process.

Burners will have a bladed diffuser or stabilizer at the end of the blast tube to provide this turbulence and give a whirling action to the air as it leaves the mouth of the blast tube, Figure 6-14. The diffus-

FIG. 6-14. Diffuser at the end of the blast tube imparts a whirling motion to the leaving air. *Courtesy Delevan Mfg. Co.*

FIG. 6-15. The amount of air delivered to the blower is controlled by an adjustable shutter. *Courtesy Wayne Home Equipment Co., Inc.*

er, by restricting the exit point, also increases the velocity of the air, giving a better mixing capability. The amount of air delivered by the combustion air blower is determined by the air control shutter.

One common type of air shutter consists of slotted rings, one of which fits over the other. The outer ring is held in place by a locking screw, Figure 6-15. By loosening this screw and rotating the outer ring, you can expose more of the open port area and increase the amount of combustion air to the burner. Rotating in the opposite direction closes the ports, reducing the amount of air. How to determine the correct amount of air will be discussed later.

TRANSFORMERS

All high pressure gun-type burners use electric ignition consisting of a step-up transformer, high tension leads, electrodes and ceramic insulators. The primary of the transformer is 120v and the secondary will be 10,000v. This voltage is produced at about 23 milliamps, (.023 amp), so it is not particularly dangerous because of the very low amperage. The coils of the transformer are protected against corrosion and other external elements by covering them with a tar-like compound. As a result, a defective or weak transformer cannot be repaired in the field, but should be replaced.

ELECTRODES

Ignition is accomplished by jumping a spark, created by the 10,000v secondary of the transformer, across a gap between a pair of electrodes located over and slightly ahead of the end of the nozzle, Figure 6-16.

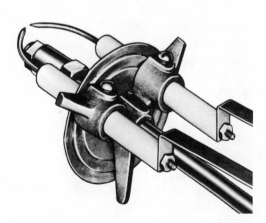

FIG. 6-16. Burner electrodes and nozzle assembly. *Courtesy Wayne Home Equipment Co., Inc.*

The gap between the electrodes and their location, both in height above the centerline of the nozzle and distance in front of it, is quite important and the manufacturer's directions should be followed in this regard. As a general guide, for residential burners, the spark gap is fairly standard and should be 3/16 in. unless the manufacturer's directions specify otherwise. Distance from the electrode tips to the centerline of the nozzle is 1/2 in. for nozzles with spray angles over 45°. This is reduced to 7/16 in. for a 45° nozzle and 3/8 in. for a 30° nozzle. The distance from the electrode tip to the nozzle center will vary according to the spray angle size,

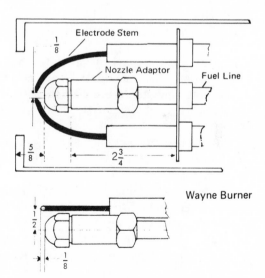

FIG. 6-17. Typical electrode spacing. *Courtesy Wayne Home Equipment Co., Inc.*

but the variation is not great. Again, a general rule, unless specifically instructed otherwise, would be 5/16 in. Also, electrodes should be at least 1/4 in. from any metal part in the burner to prevent the spark from arcing and shorting out, Figure 6-17.

The electrodes are held in special porcelain insulators, which are capable of withstanding the voltages produced by the secondary of the transformer. The insulators are held by clips or other devices, usually connected to the oil line, which is held in the blast tube by special mountings to keep it in place. The back end of the electrode will have buss bar contacts or insulated leads which complete the circuit to the secondary terminals of the transformer.

CHECKING TRANSFORMER

It should be kept in mind that the ignition transformer must be capable of producing maximum spark in order to have good ignition. Caution should be exercised in checking since there is a 10,000v potential across the contacts of the transformer.

1. Tilt the transformer open to expose the terminals and apply power to the burner.
2. Take a well-insulated screwdriver and lay it across the transformer terminals to create an arc.
3. Draw the arc as far as possible. It should be between ½ and ¾ inches in length.
4. A weak spark, that will not draw to this distance, or no spark, or a weak spark in one terminal and not the oth-

er, indicates a defective transformer. In this situation the transformer should be replaced. Another method of testing the spark emitted by the transformer is to place a small piece of newspaper in between the electrodes where the spark will hit it. The paper should ignite in 3 to 5 seconds. If not, it is an indication that there is a weak spark.

IGNITION SYSTEM

With a good transformer, the next checkpoint is the electrodes. If ignition is defective or delayed, it can create a major and dangerous problem resulting in blowback of the flame. The high voltage, produced by the transformer, will take the path of least resistance. If the insulators are cracked or damaged, the spark will jump to other metal parts in the gun-type burner. For this reason, the insulators on the electrodes should be removed and carefully examined for any cracks, breaks or darkened surfaces which would indicate leakage of electricity. No part of these electrodes should be closer than ¼ in. to other metal parts in the burner, to prevent the spark from jumping to adjacent metal parts. The electrodes should be out of the direct path of the oil spray. Oil can build up carbon across the gap, shorting out the electrodes and killing the spark. The electrodes should be examined visually to see if any carbon has built up on them. Electrode tips can be dressed with a file so that they are clean and will not resist the spark.

Any time carbon or darkened spots are detected on the insulators, it is obvious

that there is a poor, smoky fire created by the lack of draft or combustion air. This should be corrected. If the insulators do have carbon built up on them, they will no longer act as insulators. Instead, they become conductors and the possibility of a short to some other metal part will exist. Also, if there is not enough spark for good ignition and combustion, more carbon and soot are rapidly built up.

Electrodes should be checked to make sure that the ends connecting with the ignition transformer make good contact. If not, the buss bar or plate should be bent or formed to insure positive contact. Contact can be determined by feeling the resistance while the transformer is being secured in place. If any doubt exists about whether the electrodes are making good contact with the transformer, the buss bars can be coated with a tracing material and the door closed. Observing contact marks, upon reopening the door, will tell the serviceman whether or not he is making firm contact.

When replacing the burner tube and ignition assembly, be sure that the burner tube, when reassembled, does not extend into the combustion chamber, Figure 6-18. The outer edge of the burner tube should be exactly flush with the inside of the combustion chamber and not extend into it at all. On a circular combustion chamber, this means that the center of the blast tube will be exactly on the inside diameter of the combustion chamber and that the edges will be slightly back from it. On a square chamber, the entire face of the burner tube will be exactly square with the combustion chamber.

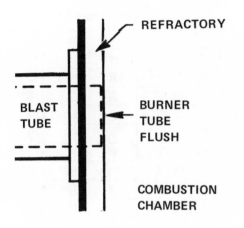

FIG. 6-18. Positioning the blast tube in combustion chamber.

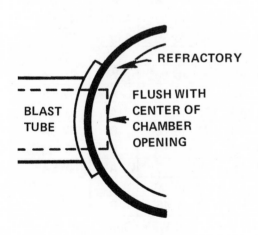

NOZZLES

The nozzle, Figure 6-19, in an oil burner has three important functions:

1. *Atomizing.* In order for oil to burn completely, it must be vaporized. This is accomplished by breaking the oil up into droplets. The droplets are very tiny, ranging from .002 to .010 in. in diameter, which are necessary for fast, quiet ignition. It is also important to establish the flame front very close to the burner head.

2. *Fuel Delivery.* The nozzle is designed to deliver a fixed amount of fuel to the combustion chamber. This is within a + or - range of 5% of rated capacity at 100 psi. This requires nozzles in a wide variety of flow rates to match the various manufacturers' furnace designs. For pumps under 5 gallons per hour, 21 different flow rates and 6 different spray angles are considered standard.

3. *Patterning.* The nozzle is expected to deliver the atomized fuel to the combustion chamber in an uniform spray pattern and at a spray angle best suited to the requirements of that particular burner. Spray patterns will range from a hollow cone to a solid cone to what is also called an all-purpose spray pattern, Figure 6-20. Common spray angles will be 30°, 45°, 60°, 70°, 80°, and 90°.

FIG. 6-19. Oil burner nozzle. *Courtesy Delevan Mfg. Co.*

FIG. 6-20. Typical nozzle spray patterns. (Left) Hollow cone. (Center) Solid cone. (Right) All-purpose cone. *Courtesy Delevan Mfg. Co.*

In a square or round combustion chamber, spray angles from 70 to 90° are normally used so that the oil pattern will cover the greatest amount of area in the chamber, Figure 6-21.

With a rectangular or long, round chamber, spray angles of 30 to 60° will be used, again to cover the maximum amount of area with the chamber, Figure 6-22. The nozzle will include a strainer because dirt or other foreign material can clog up the nozzle and cause improper combustion.

Remember that, as the outside temperature decreases, the viscosity of the oil changes. That is, it thickens and flows less freely. Outside tanks, in very cold weather, could deliver cold oil to the nozzle. Cold oil results in poor atomization. The oil may be so cold that it cannot be ignited at all. If it is ignited, it produces a long, narrow, noisy flame that burns on the back wall, rather than directly in front of the burner.

Some manufacturers recommend changing the nozzle each year because the tips are subject to considerable erosion by the oil. This changes the orifice size and the spray angle delivered into the combustion chamber. Others say, if the burner is operating satisfactorily, leave it alone. The nozzle should be checked periodically for wear and changed if the wear appears excessive. Special nozzle wrenches are available for this purpose. The new nozzle should be handled very carefully since it normally will have a body made of brass, Figure 6-23.

The several manufacturers of nozzles use somewhat different methods of measuring spray angles and patterns, but Ta-

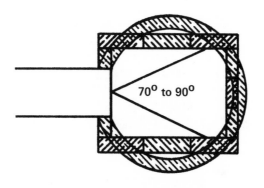

FIG. 6-21. 70 to 90° spray angles are for square or round combustion chambers.

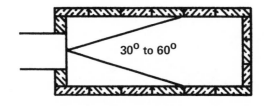

FIG. 6-22. 30° to 60° spray angles are for rectangular or cylindrical combustion chambers.

FIG. 6-23. Nozzle removed using a nozzle wrench. *Courtesy Sundstrand Corp.*

ble 6-2, developed by Delavan, shows the equivalent nozzles of the various manufacturers. Because of possible variations

TABLE 6-2. Interchange Chart Oil Burner Nozzles

Hago		Delavan	
Type	Spray Angles	Type	Spray Angles
H	30° & 45°	A	NWSA*
H	60° to 90°	A	60° to 90°
EH	30° & 45°	A	NWSA
EH	60° to 90°	A	60° to 90°
ES	30° & 45°	W	NWSA
ES	60° to 90°	W	60° to 90°
P	30° & 45°	B	NWSA
P	60° to 90°	B	60° to 90°
Unmarked Blue Vials		A	—

Monarch		Delavan	
Type	Spray Angles	Type	Spray Angles
R	30° to 60°	A or W	NWSA*
R	70° to 90°	A or W	70° to 90°
AR	45°	W	NWSA
AR	60° to 90°	W	60° to 90°
NS	30° to 60°	A	NWSA
NS	70° to 90°	A	70° to 90°
PLP	30° to 60°	B	NWSA
PLP	70° to 90°	B	70° to 90°
PL	45° & 60°	A	NWSA
PL	70° to 90°	A	70° to 90°
HV	30° & 45°	B	NWSA

*NWSA—Next Wider Spray Angles

from time to time, the chart should be considered a guide, subject to change. Most of these nozzles, however, will give equal performance.

OIL PUMP UNITS

The main function of the oil pump is to move the oil from the oil tank and deliver it, at constant operating pressure, to the nozzle. This atomizing or operating pressure will be 100 psi. If the oil tank is underground or below the level of the oil burner, then the pump must perform another function: that is, to lift the oil whatever distance it is below the level of the nozzle. Today, the fuel pump is actually a unit which includes the oil pump itself, a filter and a pressure regulating valve, all contained in one housing, Figure 6-24. The pumps are gear-type and are manufactured either in single-stage or two-stage.

Single-stage pumps have a single set of gears and are manufactured for single-pipe, gravity feed installations or for two-pipe systems under low lift conditions up to 10 in. of vacuum. The Sundstrand Model A pump can be mounted in any position. A single-stage circuit diagram is shown in Figure 6-25.

Where higher lift conditions exist, a two-stage pump is used. One set of gears

lifts the oil and the other set supplies oil under pressure to the nozzle. Under high lift, it will be used with a two-pipe system, but it can also be used on a single-pipe, gravity feed installation. In the latter application, it functions as a single-stage pump. The Sundstrand Model B may be mounted in any position except upside down. The circuit diagram is as shown in Figure 6-26.

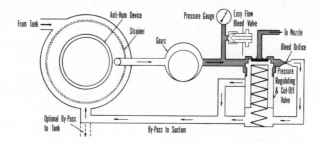

FIG. 6-25. Circuit diagram of a single-stage pump.

FIG. 6-24. Cutaway of a single-stage pump. *Courtesy Sundstrand Corp.*

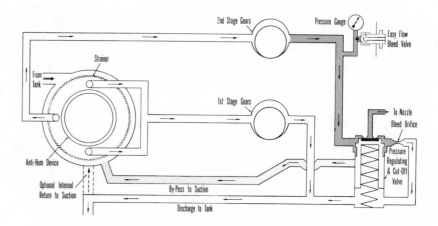

FIG. 6-26. Circuit diagram of a two-stage pump. *Courtesy Sunstrand Corp.*

Both single and two-stage pumps will have a number of outlets for the required piping connections, Figure 6-27. Note that the pump is shipped for single-pipe installation with the bypass plug removed. For two-pipe systems, the 1/16 in. bypass plug (included with the unit) must be inserted into the 1/4 in. return line port. This is done with a 3/16 in. Allen wrench.

The pump shaft is connected to the motor shaft by means of a coupling and, therefore, will always operate at a constant speed, which is the same as the motor. Oil burner pumps are designed to deliver more oil than the nozzle output capacity, in order to control the pressure at the nozzle. An additional device, which is an integral part of the pump, is the pressure regulating valve. This valve performs three functions:

1. It will not allow oil to reach the nozzle until the minimum operating pressure has been reached, usually in the neighborhood of 60 to 70 psi, since the nozzle will not atomize oil properly below that pressure.

2. It must maintain a constant operating pressure around 100 psi. This pressure provides a constant gallon-per-hour flow rate and spray pattern to the fire.

3. When the burner goes off, there must be a quick, clean cutoff of the oil supply so that there will be no after-drip of oil while the burner is off.

The final item in the unit assembly is a strainer ahead of the pump to make sure that no dirt or other foreign material enters the pump. Since the gear pump has very closely machined surfaces and tolerances, it is necessary to make sure that no foreign materials enter the pump to alter its operating capabilities. The pump and the strainer are basic service items and should be checked and cleaned with clean fuel oil or gasoline on periodic maintenance inspections, Figure 6-28. A new cover gasket should be installed after cleaning.

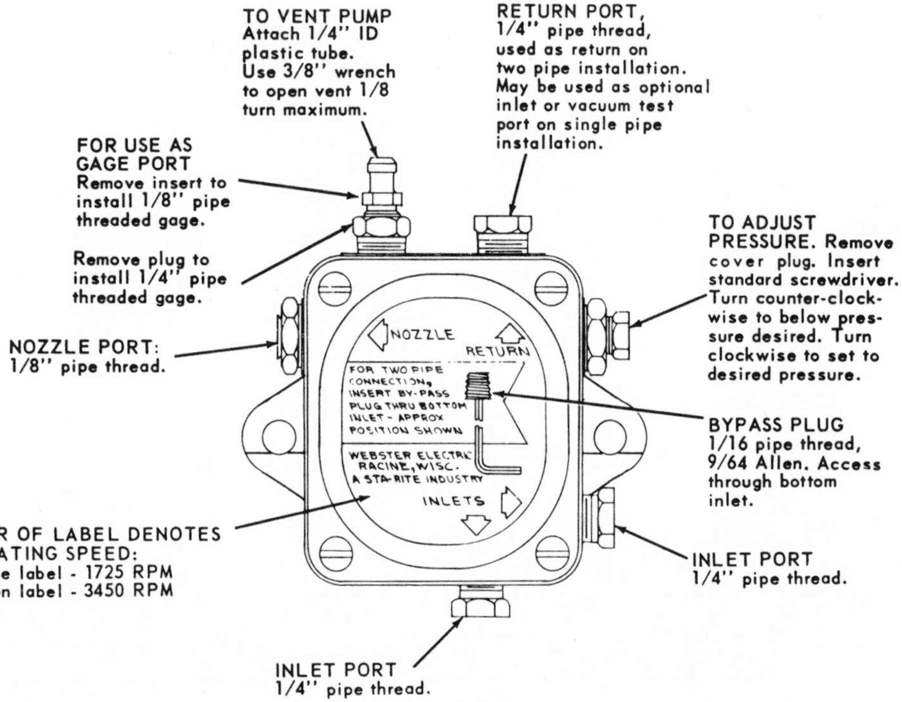

TO VENT PUMP
Attach 1/4" ID plastic tube. Use 3/8" wrench to open vent 1/8 turn maximum.

RETURN PORT, 1/4" pipe thread, used as return on two pipe installation. May be used as optional inlet or vacuum test port on single pipe installation.

FOR USE AS GAGE PORT Remove insert to install 1/8" pipe threaded gage.

Remove plug to install 1/4" pipe threaded gage.

TO ADJUST PRESSURE. Remove cover plug. Insert standard screwdriver. Turn counter-clockwise to below pressure desired. Turn clockwise to set to desired pressure.

NOZZLE PORT: 1/8" pipe thread.

NOZZLE

RETURN

FOR TWO PIPE CONNECTION, INSERT BY-PASS PLUG THRU BOTTOM INLET - APPROX POSITION SHOWN

WEBSTER ELECTRIC RACINE, WISC. A STA-RITE INDUSTRY

INLETS

BYPASS PLUG 1/16 pipe thread, 9/64 Allen. Access through bottom inlet.

COLOR OF LABEL DENOTES OPERATING SPEED:
Blue label - 1725 RPM
Green label - 3450 RPM

INLET PORT 1/4" pipe thread.

INLET PORT 1/4" pipe thread.

FIG. 6-27.   Usual pump ports. *Courtesy Webster Div., Andro Corp.*

FIG. 6-28.   Cleaning the pump strainer. *Courtesy Sundstrand Corp.*

AIR LEAKS

Air leaks are the cause of a great number of service problems. Therefore, all valves, filters, splices and other fittings should be checked for tightness. A small air leak can cause after-fire, that is, oil dribbling into the combustion chamber after the burner has shut down. This will foul the nozzle, cause odors and also can cause unusual pump noises.

All connections must be airtight to eliminate this problem. After checking the fittings for tightness, Figure 6-29, the fuel line can be purged of any trapped air in the following manner:

1. Turn off the disconnect switch to the oil furnace and remove the wires from the FF terminals (flame detection) and the TT terminals (thermostat).
2. Jumper the thermostat terminals.
3. Loosen the pump bleed valve about ½ turn and slip ¼ in. ID hose over this valve. Place the other end in a large container.
4. Close the disconnect switch, starting the burner, and jumper the FF terminals within 45 seconds so that the burner will operate continuously. *Note that the burner safety is locked out. You must be sure that fuel ignition occurs before continuing this procedure.*
5. Collect at least a pint of oil, or continue until the oil is clear and free of bubbles before stopping the bleed procedure.
6. Note that a two stage pump will bleed automatically, but if it is desirable to speed up the process, the above procedure can be followed.

FIG. 6-29. Checking tightness of fittings. *Courtesy Sundstrand Corp.*

Pump pressure and vacuum should also be checked at this time, keeping in mind that the normal operating pressure of the oil pump is 100 psi. To check this pressure:

FIG. 6-30. Checking oil pump operating pressure. *Courtesy Sundstrand Corp.*

1. Remove the plug from the pump port marked Gauge and connect a pressure gauge (0-300 psi) to this port, Figure 6-30.
2. Jumper the TT terminals of the primary control to turn the burner on. The pressure in the oil pump should rise to 100 psig, plus or minus 5 psig. If it is above or below this, the pump pressure can be adjusted by turning the adjustment screw on the pressure regulator. Turning the regulator clockwise *increases* the pressure.
3. The burner should be cycled 2 or 3 times with the TT jumper wire by removing one end, stopping the burner and then reinserting it after a wait of 2 or 3 minutes. Pump pressure should remain at 100 psig and the burner should continue to operate correctly. If not, bleeding and adjustment of the pump pressure should be repeated.
4. Remove the jumper wire, remove the gauge and reinstall the port plug. Remove the bleed hose.

Leaks in the pump and oil supply line can be detected by checking to see if a pump draws the proper vacuum. Proceed as follows:

1. Set up the pump for a single-line operation, that is, remove the bypass plug and return line fitting if necessary.
2. Install the vacuum gauge in the unused intake port. Figure 6-31.
3. Start the burner and open the bleed valve to make sure that there is oil in the pump. Stop the burner, remove the oil line from the intake elbow and cap the elbow.

FIG. 6-31. Checking vacuum draw of oil pump. *Courtesy Sundstrand Corp.*

4. Start the burner, closing the disconnect switch, and open the bleed valve until the vacuum gauge registers approximately 10 in. of vacuum.

5. Close the bleed valve and stop the burner. The vacuum gauge reading will drop slightly until the check valve closes. At this point, the vacuum gauge reading should now hold constant. If the gauge reading continues to drop after shutdown, check all plugs, nuts and caps covering the pressure adjustment screw for tightness.

6. Repeat Steps 3, 4 and 5 and, if the pump still will not hold a vacuum, it should be replaced.

7. Reinstall the oil line to the tank and run the burner to obtain a running vacuum reading. This reading should be approximately 1 in. of vacuum per each foot of lift. If the reading is substantially higher than this, there is a restriction in the line fittings and possibly the filter.

COMBUSTION SERVICE RECORD
HIGH PRESSURE GUN-TYPE BURNERS
For Use with BACHARACH Combustion Testing Instruments

Owner _John Doe_
Street _426 Maple Drive_
City _Centerville_   Phone _CE-5481-J_
If not home get key 928 Maple Drive
Occupant _____ _Same_
Street _____
City _____ Phone _____
Work Authorized ☑ by Owner ☐ by Occupant _John Doe_
Signature of person authorizing work

Order No. _4244_ Date _2-22-65_
Taken by _Smith_
Condition Reported
☐ No Fire   ☑ Insufficient Heat
☑ Excessive Oil Consumption   ☐ Odor
☐ Burner Ignites, then Goes Out   ☐ Noise
☐ Burner puffs . . . ☐ On Start; ☐ On Stop
OTHER
When Service Wanted
DATE _2-23-65_ TIME _Before Noon_
☐ PHONE FOR APPOINTMENT _____
Job Assigned to:
NAME _Ryan_   DATE _2-23-65_
Job Completed:
DATE _2-23-65_   TIME _10 25 AM_
BY _Tom Ryan_
Signature of Service Man

I—Preparing for Combustion Test
1. Open main burner switch.
2. Inspect and clean out accumulated oil in combustion chamber.
3. Advance thermostat. (5-10° F.)
4. Close remote control burner switch.
5. Make ¼″ diameter hole in flue pipe and overfire

(for BACHARACH test instruments).
6. Insert TEMPOINT thermometer (200-1000° F. range) through ¼″ diameter hole in flue pipe.
7. Open inspection port or door.
8. Adjust flame mirror.
9. Close main burner switch. (Starting burner.)

II—Combustion Test Procedure and Inspection Data

STEP	Observe—and mark with √		TEST NO. 1	2	3	4
1	FLAME IGNITION	Instant	√	√		
		Delayed				
		Doesn't Ignite				
2	FLAME COLOR	Orange		√		
	If flame shows two colors check both	Yellow	√	√		
		White	√			
		Sparks				
3	FLAME SHAPE	Uniform	√			
		Lop-sided				
4	FLAME IMPINGEMENT	At bottom				
		At sides				
		At rear				
5	ODOR	None	√			
	Near burner; observation door, draft regulator	Slight				
		Heavy				
6	NOISE	Rattle				
	Mark (x) when Excessive	Hum				
	Mark (√) if Moderate	Pulsation				
		Start				
		Running				
		After fire				
7	SOOT DEPOSIT	Flue				
	Mark (x) if Heavy;	Comb. Chamber				
	(√) if Light;	Furnace/Boiler				
	(o) if None					

STEP	Observe—and write in data		1	2	3	4
8	(Close observation door) OVERFIRE DRAFT in inches Water		.030	.020		
9	TEMPOINT READING FLUE GAS TEMP. °F. When constant temperature is reached		710	610		

STEP	Observe—and write in data		1	2	3	4
10	BASEMENT AIR TEMP. °F.		60	60		
11	NET STACK TEMP. °F. Subtract basement temp. (step 10) from flue gas temp. (step 9)		650	550		
12	FLUE DRAFT in inches water, (use same hole used for stack temp. test)		.035	.025		
13	FYRITE READING % CO_2 (use same hole used for stack temp. test)		5½	9½		
14	TRUE-SPOT SMOKE READING (use same hole used for stack temp. test)		½	1		
15	FIRE EFFICIENCY FINDER % COMBUSTION EFFICIENCY		60¼	76½		
16	(Open Main Burner Switch) FLAME CUT-OFF (Seconds) Estimate time required in seconds for flame to disappear after burner stops		2	2		
17	(Close Main Burner Switch) OIL PRESSURE (psi) Measured with oil gauge installed on pump					
18	FEED LINE SUCTION (inches) Measured with vacuum gauge installed in feed line					
19	(Open Main Burner Switch Remove Nozzle Assembly) NOZZLE (service if necessary, then reinstall)	Size—Gph				
		Type—S/H				
		Spray Angle				
20	COMBUSTION CHAMBER SIZE	Depth ″				
		Length ″				
		Width ″				
		Area sq. in.				

III—Adjustments and Repairs
Make adjustments, install replacements, and tune-up as required. Indicate, in spaces provided below, work done before repeating the tests listed under "II".

Write in "A" for "Adjust"; "C" for "Clean", "R" for "Replace". Mark "√" for other work, and describe it on the back of this sheet, if necessary.

WORK PERFORMED	BEFORE TEST NO.			
		2	3	4
BURNER AIR SHUTTER	A			
SEAL AIR LEAKS	√			
BURNER AIR BLOWER	C			
TURBULATOR				
AIR CONE				
BAROMETRIC DAMPER	A			
BURNER IGNITION— SAFETY CONTROL				
LIMIT CONTROL				
ELECTRODES				
ELECTRODE CABLE				
TRANSFORMER				
AIR FILTERS	R			
NOZZLE				
NOZZLE STRAINER				
PUMP STRAINER				
PUMP				
OIL FILTER	R			
OIL PRESSURE				
PUMP CUT OFF				
COMBUSTION CHAMBER				
BURNER POSITION				
BELT-COUPLING				
OIL LINE				
CHIMNEY REPAIRS				
FURNACE/BOILER CLEANED				

IV—Final Inspection
(a) Repeat the combustion check-ups listed under "II", and enter data in proper spaces.
(b) Check each of the following for proper setting, operation, or condition.

☑ MAIN BURNER SWITCH   ☑ THERMOSTAT
☑ BLOWER CONTROL   ☑ LIMIT CONTROL
☑ PUMP CONTROL   ☑ LUBRICATION
☐ LOW WATER CUT OFF   ☑ OIL LEAKS
☑ CIRCULATING-AIR FAN   ☑ AIR FILTERS
CONDITION OF FUEL OIL _good_
FLAME FAILURE CUT OFF TIME _120_ SEC.
IGNITION CUT OFF TIME _45_ SEC.

FIG. 6-32. Bacharach combustion testing form. *Courtesy Bachrach Instrument Co., Div. AMBAC Industries.*

COMBUSTION TESTING

For a high pressure gun-type burner to give maximum combustion efficiency, several interrelated variables must function with a specific relationship to each other. Combustion testing, to be effective, must follow an orderly and planned procedure so that the variables can be tested in the proper sequence. This sequence is a sound and accurate method of evaluating the burner performance, plus an easy method of locating the causes if the performance is not correct. By repeating the steps in the same order, it is easy to determine whether the corrective measures have proved effective.

With the proper tools, a reasonable understanding of what has to be done to improve the combustion efficiency, and a planned order of testing, the best results can be obtained for the customer. Bacharach Instrument Company of Pittsburgh, Pa. has published a *Combustion Service Testing Procedure* which will be followed in this book. As various instruments are required, their use and operation will be explained. See Figure 6-32.

The basic procedure is to run through the first 16 checks completely, recording the information, observed or measured, on a service record check. If minor adjustments will improve performance in any area, then the test can be stopped at this point. However, if more serious trouble is indicated—such as delayed ignition, sparks in the flame, bad flame shape or impingement, high setting, extremely low combustion efficiency or excessive smoke —it is advisable to complete the balance of the checks to see if the problem can be taken care of by more extensive service procedures.

PREPARATION FOR TESTING

1. Open the main burner switch. This switch will be used to start or stop the burner since it is located close to the burner. Thus it can be quickly switched off if there is a problem on initial start. The fuses in this switch can be checked at this time, particularly if the service call was one for *no* heat.
2. The combustion chamber should be inspected and any accumulated oil should be cleaned out with rags. Leaving the observation port open will usually burn out any oil absorbed in the combustion chamber.
3. The thermostat should be advanced 5 or 10° F to assure continuous burner operation while the combustion test is run.
4. If there is a remote control burner switch, this should be closed.
5. If not already available, a ¼ in. hole should be drilled in the flue between the flue collar of the furnace and the barometric regulator, at least 6 in. from the regulator. It is also necessary to have a ¼ in hole over the fire. In some cases, this might be provided by removing a bolt in the inspection door. On other units, the air louvre may be opened slightly to provide a place for insertion of the tube and the balance sealed off with paper or cardboard.
6. A stack temperature thermometer (200-1000° F range), Figure 6-33, can be inserted through the ¼ in. hole in

FIG. 6-33. Checking stack temperature with a thermometer.

the flue. The end of the thermometer stem should reach into the center third of the flue pipe diameter.

7. Open the inspection port or door and adjust the flame mirror, Figure 6-34, so that the flame can be easily observed. The main burner switch can now be closed, which will start the burner. A lighted flashlight should be available so that the oil mist can be inspected if the unit fails to ignite. If it does fail to ignite, and there is no oil mist in the chamber, oil stoppage can be suspected. If there is oil mist in the chamber, a failure in the ignition circuit is indicated.

FIG. 6-34. Using flame mirror.

COMBUSTION TEST PROCEDURES

1. If the burner ignites, note whether this ignition was instant or delayed.
2. The flame color should be checked and recorded, including the one or more colors observed. The flame will be orange, yellow, white or any one of these and may also show sparks within it. The desirable color is orange with perhaps some yellow in it.
3. Flame shape is checked to see whether it is uniform and covering a broad area of the combustion chamber. The ideal flame is a *sunflower* shape, Figure 6-35, which covers a broad area evenly. With a lopsided flame, certain areas in the chamber are not being exposed to the flame. A poor or unbalanced flame usually indicates a bad nozzle which will have to be replaced.
4. Flame impingement should be checked and, if this happens, it should be noted where it occurs, whether at the bottom, sides, or rear of the combustion chamber. If impingement occurs, a bad nozzle or incorrect size is indicated.
5. Odor should be checked near the burner, at the observation door, and at the draft regulator. No odor should be present.
6. Ignition noise should be noted. Record whether it rattles, hums, pulsates, and whether this is during start or running conditions.
7. Soot deposit should be checked at the flue combustion chamber within the furnace.

In the following steps, the actual values measured should be recorded on a check sheet and used for future reference.

8. With a draft gauge, Figure 6-36, measure overfire draft, in inches of water (w.g.) with the observation door closed. In most furnaces, the tube can be inserted through the inspection port, making sure that the metal part of the tube extends over the fire. To get an accurate reading, the tube should be sealed in the inspection port, either with a piece of sheet metal or tape, so that there will be no air leakage. Prior to taking the reading, the meter should be zeroed. The draft reading is recorded and should be in the neighborhood of .02 in. w.g. However, the draft reading is completely dependent upon the flue draft and cannot be corrected at this point.
9. Record the flue gas or stack temperature in degrees F. This temperature is taken after the burner has run for approximately 5 minutes, or until a constant temperature is reached. It is taken by inserting a thermometer in the hole drilled in the flue. This reading should be around 500° F. If it is considerably higher, it indicates that excess heat is going up the flue and that heat transfer in the heat exchanger is inefficient. Usually, as stack temperature comes down, the

FIG. 6-35.   Ideal flame is a sunflower shape. *Courtesy Delevan Mfg. Co.*

FIG. 6-36.   Draft gauge. *Courtesy Bachrach Instrument Co., Div. AMBAC Industries.*

efficiency of the furnace will go up, since there is less loss of heat from combustion.

10. Record the basement air temperature in degrees F.

11. Record the net stack temperature in degrees F. This is determined by subtracting the basement temperature, Step 10, from the flue gas temperature, Step 9.

12. The thermometer should be removed from the flue hole and replaced with a draft gauge. Figure 6-37. Most residential oil burners, with firing rates below 1.5 gallons per hour and a wide, unrestricted flue, require flue pipe draft readings of about .03 in. of water in order to maintain a draft of .02 in. of water in the fire box. The .02 in. draft in the fire box or overfire must be maintained to prevent smoke and odor returning to the basement and rumblings.

13. Take the Fyrite percentage of CO_2 reading. The same hole that was used for stack temperature and flue draft will be used for this test. The CO_2 indicator, Figure 6-38, is operated in the following manner:

a. Remove the CO_2 indicator and sampling tube from its case.

b. Place a finger over the metal end of the sampling tube and squeeze and release the aspirator bulb. The aspirator bulb should remain deflated. If it does not, the check valve on the rubber cap and the aspirator bulb is bad and needs to be replaced before the CO_2 analyzer can be used.

c. Place a finger tightly over the rubber cap end of the tube and squeeze the aspirator bulb. The bulb should

not be able to be depressed if the finger is held tightly on the rubber cap. If it can be depressed, the check valve on the metal tube end of the aspirator bulb is bad and needs to be replaced.

d. Insert the sampling tube for the CO_2 indicator into the hole in the flue pipe. Note that the metal tube has a small U-shaped end in it to hold it in place in the flue, Figure 6-39.

e. Depress the plunger on top of the CO_2 analyzer to release any flue samples that might remain from previous checks.

f. Place the CO_2 analyzer on a level surface and move the numbered scale up and down until zero is at the top of the liquid level.

g. The rubber cap attached to the end of the hose with the bulb is placed on the top of the CO_2 analyzer and pushed down to depress the plunger.

FIG. 6-37. Using draft gauge.

FIG. 6-38. CO_2 indicator. *Courtesy Bachrach Instrument Co., Div. AMBAC Industries.*

FIG. 6-39. Taking a CO_2 sample.

h. The aspirator bulb is squeezed firmly 18 times in succession.

i. On the 18th squeeze of the bulb, the plunger valve is released before releasing the aspirator bulb.

j. After removing the rubber cap from the plunger of the analyzer, the analyzer is turned over twice so that the test fluid can absorb the flue gas sample. The second time that the analyzer is turned over, it is brought back to a 45° position for a few moments to be sure that all the liquid drains down to the body of the analyzer before it is brought to the upright position.

k. By placing or holding the CO_2 indicator in an upright, level position, the level of the test liquid is read on the scale. The scale is calibrated directly in percent of CO_2.

The CO_2 reading should be between 8-10%. If it is above or below this point, proceed to the smoke test as no corrective action is taken until both variables are known.

14. The True Spot Smoke Tester, Figure 6-40, will determine whether the flame is smoking excessively and thus inclined to build up soot in the furnace. If soot does build up within the furnace, it means wasted fuel and reduced heat transfer, because soot is an excellent insulator.

This test is taken by:

a. Inserting the free end of the sampling tube into the hole in the flue pipe used in the other tests, Figure 6-41.

FIG. 6-40. True Spot smoke tester. *Courtesy Bachrach Instrument Div., AMBAC Industries.*

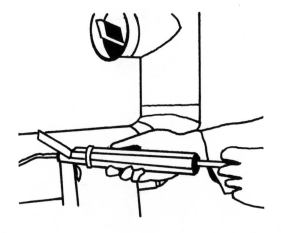

FIG. 6-41. Using smoke tester.

b. Loosen the front end of the tester slightly to allow insertion of the special filter paper into the holding slot. The tester is then retightened, using moderate pressure.

c. The smoke tester handle, which is similar to a small bicycle pump, is pulled through 10 full pump strokes. Use slow, even strokes and rest several seconds between each complete stroke.

d. Remove the filter paper from the holding slot.

e. The filter paper is then placed on the scale showing number and shade of color with a hole in the middle of each one for accurate comparison. Match the sample to the closest color on the scale, and record this number on the check sheet.

Not all burners will be equally affected by the same smoke content, so it might be wise to check the individual manufacturer's recommendations. Newer furnaces should show a No. 1 spot, but some older furnaces might accommodate a No. 2 or No. 3 spot without harmful sooting.

If the CO_2 is between 8-10% *and* the smoke spot is No. 1 or less, then the furnace is operating normally and the combustion efficiency test can be made. If the smoke is more than No. 1, increase the amount of combustion air by loosening the set screw and rotating the outer ring to expose more port area. Operate the burner for 4 or 5 minutes and recheck the CO_2.

If the CO_2 is less than 8% and the smoke spot is less than No. 1, then reduce the combustion air and recheck both.

If the CO_2 is less than 8% and the smoke is over No. 1, then suspect leakage of air at the inspection door, flue pipe or burner receiving tube. If there is no leakage at these points, the nozzle may need replacement.

15. Bacharach puts out a simple combustion effiency slide rule which directly determines combustion efficiency from the previous CO_2 and net stack temperature tests. This slide rule has both a horizontal and vertical slide insert, Figure 6-42. The horizontal slide is moved outward until the net stack temperature, determined in Step 11, shows in the window in the upper right hand corner. Then the vertical slide is moved downward until the black arrow points to the % of CO_2, as determined in Step 13, on the scale to the left of the window. The % of combustion efficiency and stack loss are indicated in the cutout in the

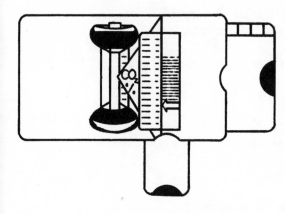

FIG. 6-42. Bachrach combustion efficiency slide rule.

center of the arrow pointing to the CO_2 reading. The reading should be greater than 75%. If it does not reach this point, then the stack temperature is too high, assuming the CO_2 and smoke readings are satisfactory. High stack temperature can be caused by too much restriction in the return air of the system, high draft or an overfired unit.

16. The main burner switch should be opened to stop the burner and the seconds required for the flame to cut off are counted. This can be done by throwing the switch and counting 1001, 1002, etc., and recording the last digit. At the same time, observe whether there is any oil dripping after the flame disappears. The flame should cut out in 45-60 seconds.

FIELD WIRING

The first step in field wiring is to run the 120v, 2-wire service from the main breaker panel in the house. The size of main breaker or fuse and the wire size are specified in the instructions that come with the unit and will be in accordance with the National Electric Code. Some areas require conduit for all or part of this service. Local codes should be checked! A disconnect switch (this can be a fused handy box with off-on switch, Figure 6-43) must be provided on or near the furnace. This is fused to the manufacturer's instructions. The two wires are connected to the line side of the box with the fuse in the hot leg (black wire). Two wires are run from the disconnect to the line voltage make-up box in the furnace. The black

FIG. 6-43. Fused handy box.

wire is connected to the wire going to the fan-limit control. The white wire is connected to the white lead from the primary and the white lead to the blower.

If the oil burner assembly is shipped separately from the furnace or if its leads are not connected, then the black wire from the burner is attached to the orange wire at the primary control. The white wire from the burner connects to the white wire at the primary.

Two low voltage wires (R and W) are run from the thermostat and connected to T and T at the primary. Cad Cell leads from the burner assembly are connected to terminals S and S (or FD and FD) at the primary. If fan control at the thermostat is desired, then G from the thermostat is connected to G at the terminal block or fan control in the furnace. This will complete the field wiring for a heating-only oil furnace.

Questions

1. What fuel is used in most domestic oil burners and what is its Btu output?
2. If the oil tank is buried outside, give five general rules for its piping.
3. Where can a single-pipe gravity system be used?
4. What is the purpose of the filter?
5. What special considerations are involved in an oil heat exchanger?
6. How might the fan and limit control in an oil furnace differ from a gas furnace?
7. What is a primary control?
8. How long should it take for the safety control to lock out?
9. If the burner locks out on safety, what should be checked next?
10. What is the purpose of the flame detector? What is its unique capability?
11. Where is the Cad Cell located? Why is the placement important?
12. What should be the normal Cad Cell resistance with the burner running?
13. Name the components of the high pressure gun-type burner.
14. Name some factors which would cause the burner motor to go off on overload.
15. What is the purpose of the combustion air blower?
16. How is the combustion air volume changed?
17. What voltage is produced by an ignition transformer?
18. How is actual ignition accomplished?
19. How are the electrodes positioned?
20. How is the ignition transformer checked?
21. Are the electrodes located in the direct path of the oil spray? Why?
22. How is the burner tube positioned in the combustion chamber?
23. What are the functions of the nozzle?
24. What are common spray angles and where are they used?
25. What is the function of an oil pump?
26. Where are single-stage pumps used?
27. Where are two-stage pumps used?
28. What is the purpose of the pressure regulating valve?
29. What problems can an air leak in the system cause?
30. What is the proper operating pump pressure?
31. Why is the combustion efficiency test important?
32. What is the proper over-draft reading?
33. What would be a normal flue gas or stack temperature?
34. What CO_2 reading is desirable?
35. What would be the maximum smoke spot for a new furnace?

One of the fastest growing products in recent years is the electric warm air furnace, Figure 7-1. Electric heat, in the form of heat elements embedded in the ceiling or floor, has been around for some time as have electric convection units and cable heat.

Due to a shortage of natural gas in some areas and a reduction in electric rates in others, the electric warm air furnace has become more popular. An electric furnace will use the same type of duct system as with gas or oil. Cabinets, filters, blowers and motor options will be the same as those discussed under gas and oil furnaces.

Houses with electric heat of any type should be as tight as possible and have better insulation than comparable houses with other energy sources. This cuts electric consumption (and thus cost) to a minimum.

Other applications of electric resistance heat can be as *additive heat* to heat pumps or in a fan coil unit where the electric heat is placed on the downstream side of the blower, Figure 7-2. The cabinet will contain a blower, electric heater and cooling coil.

Electric heat elements can be inserted into the duct system to provide additional heat for added-on rooms or rooms which may be hard to heat or a total house can be heated by elements remote from the basic air mover. Heating elements used in this manner are called duct heaters, Figure 7-3.

An electric furnace or an electric heating system differs considerably from a furnace using gas or oil. In the first place,

FIG. 7-1. Electric warm air furnace. *Courtesy The Coleman Co., Inc.*

Chapter 7
ELECTRIC HEATING

FIG. 7-2. Horizontal fan coil unit shows electric resistance heater on the downstream side of the blower. *Courtesy International Environment Corp.*

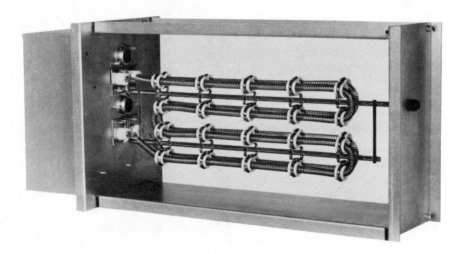

FIG. 7-3. Electric duct heater. *Courtesy Gould Inc.*

there is no combustion in an electric furnace and therefore no need for a heat exchanger. The heat is produced by a resistance heating element consisting of wound 20-gauge nichrome wire looped in a serpentine fashion along a rack, Figure 7-4. These elements are placed in the airstream and transfer their heat directly to the air without any intermediate heat exchanger. An electric furnace does not need any access spaces in the cabinet to provide combustion air, nor does it need a flue to remove the products of combustion. An electric furnace is considered to be 100% efficient, therefore there is no difference between the heat input and output of this type of furnace. For each kilowatt (1000 watts) of electricity put in, 3415 Btu's of heat energy will be produced.

The electric heat elements may be used singly or in multiples to produce the required heat output. A single element will have between a 4.5 and 6 kw input, or between 15,000 and 20,000 Btu's of heat output. Where multiple heating elements are used, they will usually be step-started, which means that the second element will not come on until the first element is producing heat. There are as many steps as there are elements, and *stepping* will continue until all elements are producing heat. The reason for this is to avoid throwing the entire electrical load on the line at one time and overloading the circuits under starting conditions. The elements are not position-sensitive, so the same ones can be used in an up-flo, down-flo or horizontal furnace. Each element will be on a separate branch circuit with fuses in both legs of the line.

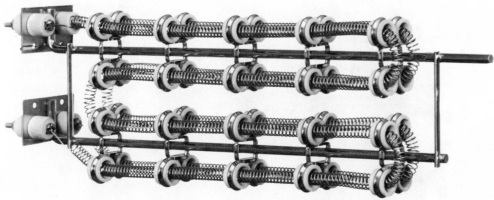

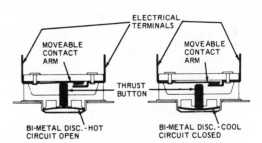

FIG. 7-4. Electric heating element. *Courtesy Gould Inc.*

Elements can be checked for proper operation by pulling out the bracket and visually checking for any broken or loose wires. If there are no visible defects, a continuity check can be made with an ohmmeter or test light.

Multiple elements will start in sequence with a slight time delay between each element coming on the line. This can be easily checked with an ammeter attached to the main load wires of the circuit. Set the ammeter on the highest current range. As each element comes on, there will be a very definite jump in current indicated on the ammeter. By counting the surges, as the elements come on the line, it can be determined whether or not all elements have sequenced on in proper order.

If all of the elements do not come on in sequence, there is a possibility that one stage may be bad or, if the unit has an outdoor thermostat, this may be holding off the last one or two elements. If this happens, the outdoor thermostat can be reset above the outside temperature. This should allow the final elements to come on the line in sequence.

LIMIT CONTROLS

An electric furnace will have a limit control for *each* element. Its function is to cut off the element if the surrounding air temperature becomes too high, just as in a gas or oil furnace. However, it is a different type of limit than is used on other furnaces. The limit control is a Klixon type of temperature-activated switch, Figure 7-5. Cutout settings may vary, but most will be between 160-170° F. The setting is nonadjustable and is determined by the manufacturer. The limit will automatically reset if the ambient temperature decreases.

FIG. 7-5. Klixon-type limit control. Photo *courtesy Gould Inc.*

The limit is mounted in the front bracket plate and senses both the temperature in the compartment and the radiant effect of the element. Each element has its own limit which will interrupt power to that element only. It will not have any effect on the other elements. Because each element has a limit monitoring its own temperature, there is no need for a secondary limit for horizontal or down-flo units as in oil or gas furnaces.

The limit can be checked by either disconnecting the blower motor or by blocking the return air. A thermometer is placed in the cabinet and the limit should open, cutting off power, when the temperature reaches 170° F (or the designed cutoff point). There should not be much variance in this cutoff point. If the temperature rises 10-15° F over the cutoff point without opening, shut off the furnace and replace the limit.

THERMAL FUSE

Another safety device on an electric furnace is called a thermal fuse or fusible link, Figure 7-6. This is set to open at a temperature that is considerably higher than the limit, 300-310° F. The purpose of the thermal fuse is to act as a backup limit. In the event that the limit does not open on an excessive temperature rise, the thermal fuse will melt and open the circuit. The thermal fuse is in series with the limit and the heating element. If the thermal fuse opens, or melts, it will not automatically reset like the limit. Some must be manually reset. In others, the thermal fuse is replaced. Earlier electric furnaces required the replacement of the entire element if the thermal fuse opened.

CAUTION While newer units, with replaceable fuses, allow this check, there is no valid service check for proper operation of the limit control on older units with nonreplaceable thermal fuse. If the blower belts are removed or direct drive leads disconnected, as in a gas furnace limit check, the temperature will not only exceed the limit control point, but will

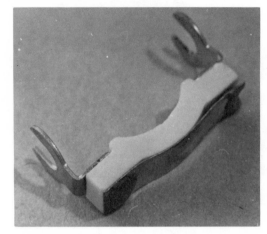

FIG. 7-6.   Fusible link. *Courtesy Gould Inc.*

FIG. 7-7.   Fan and sequence control. *Courtesy Cam-Stat, Inc.*

burn out the fusible plug, requiring changing of the entire element. A thermometer should be used to check the limit. If the limit is faulty, the furnace can be turned off before the thermal fuse melts as there usually is a wide differential (130-140° F) between them.

FAN AND SEQUENCE CONTROL

Since an electric furnace heats up rather rapidly, it is desirable to bring on the blower at the same time or slightly ahead of the first heater element. Bringing them on together is done by using a sequencer, Figure 7-7, which energizes the blower and the first element simultaneously or with a time delay of 30 seconds (this delay period can vary according to the control used). 20 seconds later, the next element is energized and so forth. Shutoff occurs (after a 3-minute delay) in the reverse order: that is, the last element to come on cuts off first. The blower circuit will be the last one de-energized, assuring blower operation until all elements are off.

To bring the blower on first, two time delay relays are used, Figure 7-8. The first will have a time delay of approximately 15 seconds after being energized and then will turn on the blower. The second relay

FIG. 7-8.   Time delay relay. *Courtesy Cam-Stat, Inc.*

will be energized at the same time but will have a delay of 20 to 25 seconds before energizing the first heating element. Additional relays, with longer time delays, can be used to bring on additional elements. This system will also shut down in reverse order, keeping the blower operating until all elements are off.

Another sequencing system will include a set of auxiliary, low voltage contacts with each delay relay. When relay No. 1 is energized, the auxiliary contacts also close, energizing the heater in relay No. 2. When this relay is energized, it brings on element No. 2 plus energizing the heater in relay No. 3 and so forth for as many elements as are in the system.

To check operation, set the thermostat to call for heat. Place the fan selector switch on automatic. Place the voltmeter leads on terminals C and W_1 on the low voltage terminal block. The meter should show 24v. Place the leads on each of the high voltage terminals. A voltage reading here indicates that the contacts are open. They should make within 30 seconds (or whatever the time delay period is) and show no voltage. Turn off unit and, when the contacts open, a voltage reading will show on the meter.

HEAT ANTICIPATION

Heat anticipation for an electric furnace is checked in exactly the same manner as heat anticipation for a gas or oil furnace. One major difference is that the heat anticipation setting usually will be at a considerably lower point (.25 to .30 amps) than it was for a gas or oil furnace, but it can run over 1 amp where several

elements are staged with the heaters in series.

To set the anticipator, first move the thermostat to its lowest setting. With a single-stage thermostat, one clip of the amperage multiplier, described earlier, should be attached to the heat anticipator lever and the other to the common terminal of the wires to the mercury bulbs. After a minute or two of operation, a reading can be taken and converted into actual amperage by dividing by 10. The heat anticipator lever is then set at this point, Figure 7-9.

With a two-stage thermostat, the first stage is checked as if it were a single-stage thermostat and the heat anticipator lever set to the reading obtained. When checking the second stage, a jumper wire is

placed across the previously used common terminal and the first-stage anticipator lever. Then the amperage multiplier is attached to the second-stage heat anticipator lever and the other side to the common terminal with the wires running to the mercury bulbs, Figure 7-10. After a minute or two of operation, the amperage reading can be taken and divided by 10 to get the actual amperage. This amperage is then set on the second-stage heat anticipator.

TEMPERATURE RISE

Because an electric furnace converts 100% of its energy output to heat energy, with no stack losses, temperature rise across the electric heat elements can be

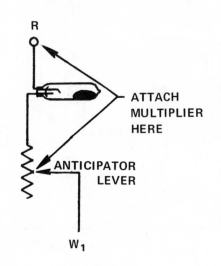

FIG. 7-9.   Single-stage heat anticipator.

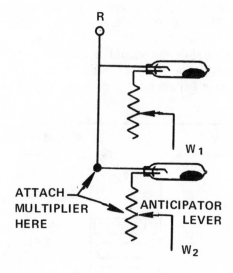

FIG. 7-10.   Two-stage heat anticipator.

less than that considered normal in a gas or oil furnace. In a gas or oil furnace, the temperature rise across the heat exchanger is between 80 and 100° and is normally held at about 90° F.

The temperature rise across an electric furnace is measured with one thermometer in the return air plenum, close to the furnace, and a second thermometer in the supply plenum, just out of sight of the electric element, Figure 7-11. After the furnace is in operation for several minutes, each thermometer should be read and the difference between them noted. For an electric furnace, this difference should be between 50 and 70° F. (Some furnaces can be as high as 85°. Check manufacturer's instructions.) If the temperature difference is less than 50°, the blower is running too fast and should be slowed down, as described earlier under Gas Heating. If the temperature difference is more than 70°, the blower is running too slow and should be speeded up.

As can be seen, the maintenance of an electric furnace is quite simple. Since there is no combustion, an electric furnace has no major moving parts, other than the blower, and no flue or vent connections. Major maintenance requires only a check of all the fuses, and electrical circuits, for tightness, plus basic blower and filter maintenance as described under Gas Heating.

FIELD WIRING

An electric furnace will require either 240v or 208v single phase service. This will be a three-wire line voltage supply run from the main panel in the house to the furnace. The disconnect and the fuse in the main supply will be sized in accordance with the manufacturer's instructions, which come with the unit. The wire size is also specified in the instructions in accordance with the National Electric Code.

One circuit is permissible to supply power from the main supply to the disconnect. The disconnect switch, Figure 7-12, must be provided at the furnace and must be located within sight of the furnace or be

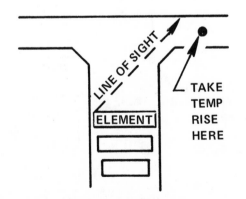

FIG. 7-11.   Checking temperature rise.

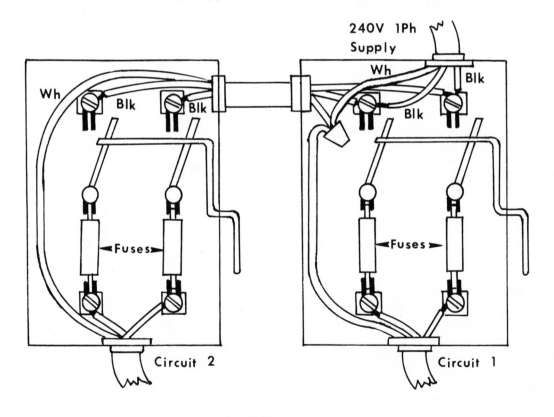

FIG. 7-12.   Field wiring for a pair of disconnects.

capable of being locked in the open position.

If the furnace does not draw over 48 amps, then only one disconnect is required. This disconnect cannot be fused for more than 60 amps. If the furnace requires greater amperage, then additional disconnects have to be supplied, one for each circuit, not exceeding 48 amps. These

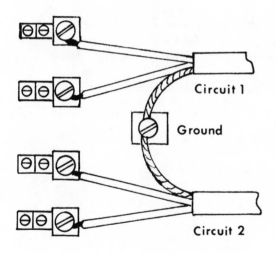

FIG. 7-13. Grounding circuits in the control box.

circuits must be enclosed in conduit and watertight strain reliefs or connectors are required at both ends. Fuses are required for each of the hot or black wires in the disconnect. These fuses can be either a standard cartridge-type fuse or a circuit breaker.

The National Electric Code also requires that these circuits be grounded at the furnace and most furnaces will provide a ground connection in the control box adjacent to the high voltage terminals, Figure 7-13. The white wire is grounded at this point and the two black wires connected to the line voltage terminals in the furnace.

Two low voltage wires (three if there is fan control and four with two-stage heat plus fan control) are run from the thermostat to the furnace. Most modern furnaces will have a terminal block or fan control center marked in the same manner as the thermostat for the low voltage leads. Thus R, W_1, W_2 and G, from the thermostat, connect to the R, W_1, W_2, and G terminals in the furnace. This will complete the field wiring.

FURNACE MAINTENANCE

As a review of the various procedures discussed up to this point, the basic maintenance procedures on all types of furnaces should always include the following:

1. All line voltage wiring and connections should be checked for tightness, good insulation and proper contact.
2. All low voltage wiring and connections should be checked for tightness, good insulation and good contact.
3. Filters should be changed every 6 months or oftener if they are dirty.
4. Blower belts should be checked for cracks, wear or excessive slippage.
5. Pulleys and drive set screws should be checked for tightness.
6. Pulleys and drives should be checked for proper alignment and belt tension adjusted if necessary.
7. All motors and bearings should be lubricated per the manufacturer's instructions and checked to see that they are free and not obstructed in any way.
8. The limit control should be checked for proper operation.
9. Temperature rise through the furnace should be checked to see that the blowers are running at the proper speeds and the furnace is functioning as efficiently as possible.
10. Thermostats should be checked for level and heat anticipation checked for proper setting.
11. Burners, heat exchangers and cabinet compartments should be checked for cleanliness and proper operation.
12. Safety devices such as thermocouples, gas valve, safety dropouts, Cad Cells and so forth should be checked for proper operation.

If all these checks are made periodically, they will go a long way towards preventing major service problems. They will also eliminate the majority of minor service problems, since potential problems will be found and corrected before they result in furnace down time.

Questions

1. How does an electric furnace differ from a gas or oil furnace?
2. How many Btu's will be produced with an input of 1000 watts?
3. When multiple elements are used, why are they started in sequence rather than coming on all at once?
4. How is the limit control on an electric furnace different than that on a gas furnace?
5. Is there a secondary limit for horizontal units? Why?
6. What is the purpose of the thermal fuse?
7. When is the blower started on a call for heat in an electric furnace? Why?
8. What is the temperature rise for an electric furnace?
9. Name ten basic maintenance items for all types of furnaces.

In order to have a complete, *total comfort* system which maintains uniform temperatures year-round, it is necessary to have cooling equipment to remove the heat gained during the summer months. Therefore, the well-rounded serviceman must know both heating and cooling. This segment limits itself to air-cooled air conditioning equipment, which is by far the most popular and widely used in the United States.

The previous chapters were concerned with adding heat to the space in order to make up the heat loss due to the difference in temperature between the inside space and the outside air. Cooling, or air conditioning, concerns itself with exactly the reverse of this process: that is, to *remove* heat from the space, heat which has flowed from the warmer outside air to the cooler inside space. Thus, our previous definition of heat still holds true in every regard. It is only the process of transferring heat, from one place to another, that is being reversed.

Chapter 8
COOLING FUNDAMENTALS

AIR CONDITIONING

Air conditioning cannot be defined as merely reducing the temperature of the inside air, since the cooling system has several other functions. Just as in the heating season, air must be circulated throughout the house. Therefore, the same furnace air mover is used in cooling as in heating. The air must also be clean, so the filters are equally important during both the cooling and heating seasons. Humidity must also be controlled, but humidity is a different problem with cooling than with heating. In heating, it is usually necessary to add moisture, since the cold outside air will not contain a great deal of moisture. When this air is heated, the relative humidity decreases and additional moisture must be added. In summer, the opposite is true, since warm air will hold considerably more moisture. The relative humidity rises as the temperature is reduced, so the moisture content of the air, or humidity, must be reduced. Finally, the temperature of the air must be reduced to maintain comfort conditions. Just as in heating, a temperature range of 70-75° F, with a relative humidity of 40 to 60%, is about the comfort range for the majority of people, Figure 8-1.

Earlier discussions of heat established that heat is energy moving from a warm body to a colder body, and that adding heat energy to a substance will change the

kinetic energy of that substance. Heat energy has neither weight nor dimension, but its presence can be detected and measured by its effect on other substances. Thermometers are used to measure temperature, which is the *intensity* of heat not the *quantity*. Measurements in the United States are normally taken on the Fahrenheit scale, which establishes a point at 32°, where water will freeze, and 212°, where water will boil at sea level. All of the points on a thermometer are indications of *sensible heat* since they can be measured and read directly. A change in sensible heat can also be sensed by touch.

The quantity of heat is measured in British thermal units or Btu's. The Btu was defined as the amount of heat necessary to raise one lb of water one degree F.

Thus, one Btu will raise the temperature of one pound of water, one degree (at sea level). In cooling, it will be necessary to remove heat. Therefore, a cooling Btu might be defined as the amount of heat which has to be removed to *reduce* the temperature. Note that heating raises the temperature and cooling reduces the temperature, but in both cases the measurement of heat quantity required to do this is in Btu's. It therefore takes one Btu to raise or lower the temperature (sensible heat) of one pound of water one degree F, or 180 Btu's for a change in temperature from 32° to 212° F.

LATENT HEAT

A new term, which was not discussed earlier, is *latent heat*. In order to change a pan of water at 32° to ice at 32°, it is

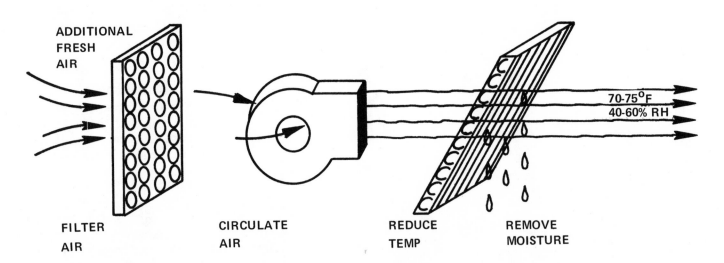

ADDITIONAL
FRESH
AIR

70-75°F
40-60% RH

FILTER
AIR

CIRCULATE
AIR

REDUCE
TEMP

REMOVE
MOISTURE

FIG. 8-1.   In addition to being cooled, conditioned air is filtered and dehumidified to the desired comfort level.

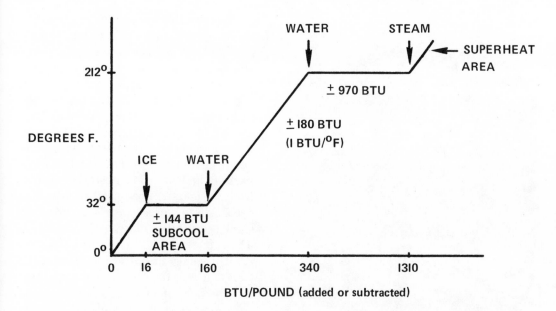

FIG. 8-2. Latent and sensible heat content of water. Horizontal lines represent latent heat since the heat change occurs without a change in temperature.

necessary to remove additional heat (reducing its kinetic energy) before the water will change from a liquid to a solid. This additional heat is called latent heat, because removing it does not change the *temperature* of the water at all. In the case of water, 144 Btu's of heat energy must be removed before the water will change from liquid to solid, *but the temperature will remain at 32°* . This is the reason why latent heat is sometimes called *hidden heat,* since it cannot be directly measured with a thermometer. The change from liquid to solid, or solid to liquid, is called *change of state.*

This *change of state* also occurs when a substance changes from a liquid to a vapor, or from a vapor to a liquid. A pan of water at 212° requires the addition of another 970 Btu's before it will change state into a vapor (steam). Water can exist either as a liquid or a gas at a temperature of 212° depending upon its total heat content. Figure 8-2 is a summary of the heat required to change the state of water from solid to liquid to vapor.

Pressure also has a bearing on this change of state. At atmospheric pressure, in any open container, the temperature of water cannot be raised above 212° F because it will change state and boil away. However, if the water is held at a constant pressure in a confined container, then its temperature can be raised above 212° without the majority of the water changing into steam.

REFRIGERANTS

A mechanical refrigeration system depends totally on the process of evaporation (the change of state from liquid to gas) and condensation (the change of state from gas to liquid) in order to function efficiently. The importance of this change of state can be illustrated by recalling the number of Btu's required in changing the state of water from a liquid to a gas. It only required 180 Btu's to raise the temperature of a pound of water from 32° to 212° , but it required 970 Btu's to change state from water to gas, or almost 5½ times the heat energy. An efficient air conditioning system must have the capability of readily absorbing and rejecting large amounts of heat at the normal operating temperatures of the equipment. Also it must be possible to continuously repeat the process of evaporation and condensation with the same substance.

While water could be used for this purpose, it boils at temperatures too high for ordinary cooling and it freezes at temperatures too high for the low temperature conditions. Therefore, some very special liquids were developed to accomplish efficient cooling. These refrigerants also had to have certain other characteristics:
1. They had to be small in volume but high density.
2. Operate at low differences in pressure.
3. Be nonflammable.
4. Be nonexplosive in either gas or liquid form.
5. Be noncorrosive.
6. Be nontoxic.
7. Be able to carry oil in solution.
8. Have a high resistance to electricity.

Most early liquids used for refrigeration—ammonia, sulphur dioxide, methyl chloride, propane, methane and others—all were lacking at least one of these qualities. These liquids have been almost completely replaced by the specialized refrigerants which were developed to eliminate the undesirable qualities of earlier refrigerants.

One of the first refrigerants developed specifically for the air conditioning and refrigeration industry was Refrigerant 12. It is widely used in household refrigerators, commercial refrigeration and window coolers, but is not used extensively in residential air conditioning. Refrigerant nomenclature is reduced from Refrigerant 12 to R-12, and all refrigerants come in color-coded drums. R-12, for instance, comes in a white container regardless of the manufacturer.

A later development was Refrigerant 22, or R-22, which is similar in many characteristics to R-12. It does have a much higher latent heat of evaporation, 86 Btu's per pound, which allows the absorption of greater heat content, and it has a higher density and lower specific volume than R-12. Therefore, more refrigerant can be pumped through the same size compressor or an equivalent amount of refrigerant can be pumped through a smaller compressor. This has resulted in smaller compressors with greater capacities than were previously available. R-22 is nontoxic, is a clear liquid, is odorless, will not burn or explode, and can be used on many varied applications. It comes in a light green container, Figure 8-3. Containers come as 125 lb cylinders, or as 50, 20 and 10 lb alumi-

FIG. 8-3.   R-22 packaged in a 50-lb disposable container. *Courtesy E.I. du Pont de Nemours & Co., Inc.*

num disposable cans. There are, however, several safety precautions in working with *any* refrigerant that should be very carefully followed.

1. Refrigerants are kept under high pressure and their containers should not be heated or left in the sun. External heat on the container will raise the pressure inside, causing the refrigerant to expand. If the pressure becomes excessive, the container can burst.

2. R-22 boils at a very low temperature, approximately -41° F. If released to atmospheric pressure, it will, of course, turn into a gas. Since it boils at an extremely low temperature, it can cause severe burns if it touches any part of the body. Be extremely careful that refrigerant does not get into the eyes. *Safety glasses should be worn in working with any refrigerant.*

3. The refrigerant will not burn. However, if it comes in contact with an open flame, it will form phosgene gas. This is a very poisonous gas which is both colorless and odorless, and was used in chemical warfare. Thus, it is extremely dangerous.

4. Refrigerants are processed to be free of water down to about 10 ppm (parts per million). Water in a system, when it comes in contact with the heat from the compressor, breaks down the refrigerant into hydroflouric and hydrochloric acids. These acids pick up copper from the tubing in the system and deposit it on the steel cylinder walls of the compressor. The compressor piston has extremely small tolerances and a very small amount of copper plating can seize the compressor. Precautions should be taken during installation so that moisture is not introduced and all systems should include a drier to help remove any moisture that might be in the system and not removed during evacuation.

SATURATED VAPORS

The pressure of any *saturated vapor* corresponds to its temperature. This fact is one of the fundamental tools used by servicemen in diagnosing the operating conditions of an air conditioning system. If you know the temperature of a saturated vapor, you also know its pressure. Conversely, if you know the pressure you also know the temperature, Table 8-1. Ordinarily, servicemen find it more accurate to apply service gauges to read pressure and thereby determine the temperature of the vapor.

By definition, a saturated vapor exists in the presence of its liquid. Another way of putting it is to say that a saturated vapor is in the two-phase condition. This is an absolutely accurate expression because, in a saturated condition, vapor is constantly condensing into a liquid, to be replaced by liquid that is evaporating into vapor. As a result, there is an equilibrium in the percent of vapor at a given pressure.

There can be two other conditions of this liquid-vapor mixture: the vapor may be superheated and the liquid may be subcooled.

SUPERHEATED VAPOR

With the application of sufficient heat, all the liquid refrigerant may be evaporated. The refrigerant is now at 100% saturation, indicating that no more liquid is available. If additional sensible heat is applied, the vapor is said to be *superheated* and the temperature of the vapor will exceed the saturated temperature at the same pressure. This condition will be used later to set the metering valve

TABLE 8-1.
Saturated Pressure/Temperature Chart

R-22		WATER
Temp °F	Pressure psig	Boils at: psig
300		52 lbs
250		15
212		ZERO
175		*16" Vacuum*
150	382 lbs	*22"*
130	297	*25"*
120	260	*26"*
110	226	*27"*
100	196	*28"*
90	168	*28.5"*
80	144	*29"*
70	121	*29.2"*
60	102	*29.4"*
50	84	*29.6"*
45	76	*29.7"*
40	69	*29.8"*
35	62	
30	55	
25	49	
20	43	
15	38	
10	33	
5	28	
ZERO	24	
—20	10	
—40	ZERO	

so as to protect the compressor.

SUBCOOLED LIQUID

If the saturated refrigerant is sufficiently cooled by the ambient air, all of the vapor will condense. The refrigerant is now at zero saturation, indicating that no more vapor is available. If the liquid is cooled to a still lower temperature, by the removal of sensible heat, the liquid is said to be *subcooled* and the temperature will be below the saturated temperature at the same pressure. Normally, the liquid line (description to follow) is subcooled for some portion of its length as it leaves the condenser.

ENTHALPY

Enthalpy is defined as the total heat content, both latent and sensible, of the refrigerant and is expressed in Btu's per pound. Enthalpy is a convenient way of measuring the heat absorbed from the living space by the refrigerant and, again, the heat rejected to the outside atmosphere. Since the enthalpy values have already been calculated, they are usually read from a Pressure-Enthalpy diagram for the particular refrigerant.

REFRIGERATION CYCLE

Up to this point, cooling has been discussed as removing heat from the airstream, but how this is accomplished and in what quantities was not covered. Now, let us cover the refrigeration cycle to determine what happens at various points in the system in terms of temperature, pressure, enthalpy and state.

Earlier, a definition of latent heat was

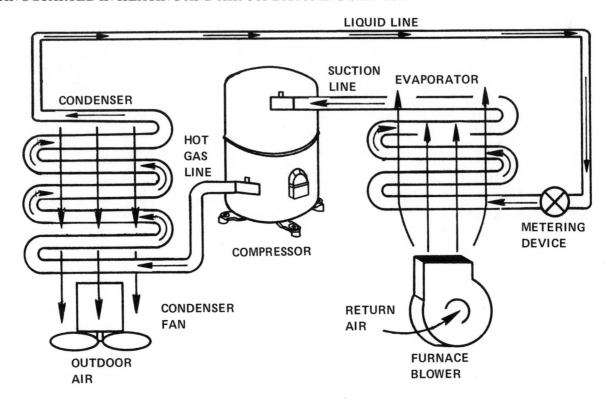

FIG. 8-4. Flow of refrigerant through an air conditioning system.

presented and this concept is fundamental to the complete refrigeration cycle. The additional heat required to cause a substance to change state, in this case from a liquid to a gas and back again, is considerably greater than that required to raise its temperature one degree.

Whenever a system calls for cooling, three or four things happen simultaneously:

1. First the compressor is energized and begins operation, pumping vaporized refrigerant out of the evaporator,

compressing it and sending it to the condenser, Figure 8-4. This immediately creates a difference in pressure between the high and low sides.

2. The condenser fan is energized and begins blowing outside air across the condenser coil so that the heat within the refrigerant vapor will be dissipated to the outside air.

3. The metering device, whether it be an expansion valve or a capillary tube, will begin passing liquid refrigerant into the evaporator so that it can begin

to pick up heat from the airstream around the evaporator.

4. If the system is not on continuous blower operation, the evaporator blower will come on, passing the warm air from the space across the face of the evaporator, so that the heat within the air can be picked up by the refrigerant.

Under theoretical conditions, the refrigerant at the metering device is in a liquid form at a 114° temperature (having been subcooled 16° F after leaving the condenser), approximately 299 psig, and has a heat content or enthalpy of 45 Btu's, Figure 8-5. The liquid refrigerant passes through the metering device into the low side of the system, and immediately expands and cools part of the refrigerant. As the liquid passes through the evaporator coil, it picks up heat from the airstream around it and begins changing to a vapor. At the exit of the evaporator, the vapor has a temperature of approximately 45° and a pressure of about 77 psig. Before entering the compressor, it is superheated 10° to 55° F, but the pressure remains constant at 77 psig. Its enthalpy or heat content, however, will have increased to 100 Btu's, having picked up 64 Btu's of latent heat from the room air and one Btu of sensible heat due to superheat.

The vapor is then pumped into the compressor shell where it passes over the motor and picks up additional heat from the motor, amounting to approximately 24 more Btu's, giving the vapor a total additional heat content of 89 Btu's. As it passes into the compressor cylinder, the vapor is compressed. At this time, its temperature will be raised to about 230° and the pressure, during the short time that it is in the compressor cylinder, will be raised considerably. Its heat content, or enthalpy, after the heat of compression, will amount to about 134 Btu's.

The vapor passes from the compressor discharge port through the discharge line and into the top row of the condenser. At this point, its temperature is 130°, it is at about 299 psig and after the first one or two rows in the condenser, will lose about 20 Btu's, so its heat content will be 114

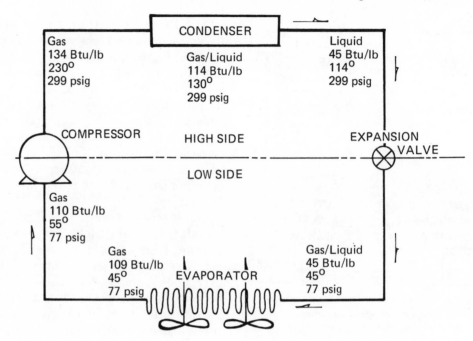

FIG. 8-5. Pressures, temperatures, and total heat content (enthalpy) at various points in the refrigeration system.

Btu's per pound.

As it passes through the remaining rows in the condenser, the vapor loses more heat to the outside air and changes state from a gas back into a liquid. As it gets to the bottom row of the condenser, all of the refrigerant will have changed to a liquid. It will be at 130° and 299 psig, with a heat content or enthalpy of 51 Btu's. The loss of 83 Btu's to the outside air is all latent heat due to its change of state from a gas to a liquid.

Upon leaving the condenser, the liquid is subcooled, thus eliminating another 6 Btu's. So, as it approaches the metering device for another circuit through the system, the liquid will be back to its original conditions: 114° temperature, 299 psig, with a heat content of 45 Btu's.

Figure 8-6 is a summary of the temperature, pressure and state of the refrigerant at various points in the system. Some systems will have better or poorer operating efficiencies. Therefore, the values given in this example will vary according to the efficiency of the system. It was also noted that the operating temperatures and pressures will change in a sys-

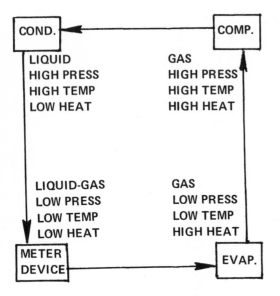

FIG. 8-6. Summary of ideal conditions at the exit of each component in a refrigerant system.

tem, depending upon the outdoor air conditions as well as the indoor temperature conditions. However, the relationship between the outdoor and indoor conditions and the enthalpy or total heat transfer of the system will remain pretty much the same.

The Btu's, temperatures and pressures indicated in this system are common for a system operating under ideal conditions. The principles, even though some of the values might change, will be the same for all types of systems.

The refrigeration cycle can be plotted on the Pressure-Enthalpy diagram as shown in Figure 8-7. Subcooled liquid, at Point A, begins losing pressure as it goes through the metering valve, located at the point where the vertical liquid line meets the saturation curve. As it leaves the metering point, some of the liquid flashes into vapor and cools the liquid entering the evaporator at Point B. Notice that there is additional reduction in pressure from the metering point to Point B, but no change in enthalpy.

As it passes from Point B to C, the remaining liquid picks up heat and changes from a liquid to a gas, but does not increase in pressure. Enthalpy, however, does increase. Superheat is added between Point C, where the vapor passes the saturation curve, and Point D. About 1 Btu is added to the enthalpy, due to superheat.

As it passes through the compressor, Point D to E, the temperature and the pressure are markedly increased, as is the enthalpy, due to the heat of compression. Line E-F indicates that the vapor must be de-superheated, within the condenser, before it attains a saturated condition and begins condensing. Line F-G represents the change from vapor to liquid within the condenser. Line G-A represents subcooling within the liquid line or capillary tube, prior to flow through the metering device.

Note that the pressure remains essentially constant as the refrigerant passes through the evaporator, but that its temperature is increased beyond the saturation point, due to superheat, before it enters the compressor.

The pressure likewise remains constant as the refrigerant enters the condenser as a vapor and leaves as a liquid. While the temperature is constant through the condenser, it is reduced as the liquid is subcooled before entering the metering valve.

The change in enthalpy, that is the heat content, as the refrigerant passes through the evaporator is almost all latent heat since the temperature does not change appreciably. At 45° temperature (normal evaporator operating temperature) and in a liquid state, one pound of R-22 contains 23 Btu's and in the vapor state it contains 109 Btu's. The heat absorbed by vaporization of R-22 therefore is 86 Btu's. This means it takes 86 Btu's of latent heat to convert R-22 from a liquid to a gas.

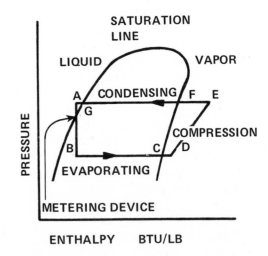

FIG. 8-7. Simplified Pressure-Enthalpy curve with refrigeration system pressures and enthalpies superimposed.

REFRIGERATION OILS

Special refrigeration grade oils which are de-waxed and dehydrated to correspond with compressor and refrigerant standards are the only oils approved for modern compressors. Some brands are: Suniso 3G, 3 GD and 4G; Texaco Capella B and Ansul 150. These are available from refrigeration wholesalers. Only small amounts should be purchased at any one time, since once opened, the oil will pick up moisture and can contaminate the system. Oils can be obtained in aerosol cans which keep them sealed off from the atmosphere and these are handy when small amounts are needed.

Questions

1. What factors are involved in air conditioning?
2. What is latent heat?
3. How much heat is removed to change water from a liquid to a solid?
4. How much heat is required to change water from a liquid to a gas?
5. What is the boiling point of R-22 at atmospheric pressure?
6. What is enthalpy?
7. What is saturation point?
8. What is superheated vapor?
9. What is subcooled liquid?
10. Name six basic characteristics of a refrigerant.
11. Why is R-22 a better product than R-12 for central air conditioning?
12. Name four basic safety factors in working with refrigerants.
13. What happens to the refrigerant as it passes through the evaporator?
14. What happens to the refrigerant as it goes through the compressor?
15. What happens to the refrigerant as it passes through the condenser?
16. What happens to the refrigerant as it goes through the metering device?
17. Name the four things that happen when a system calls for cooling.
18. What is the normal operating pressure and temperature at the entrance to the metering device?
19. What would be the normal temperatures and pressures at the exit of the evaporator?
20. What is the state of the refrigerant at the exit of the condenser?
21. What is the state of the refrigerant at the exit of the evaporator?

Chapter 9
COOLING COMPONENTS

Courtesy Copeland Corp.

Courtesy Tecumseh Products Co.

Courtesy Tecumseh Products Co.

Courtesy Copeland Corp.

The compressor is the heart of the refrigeration system and will be the most expensive component. The compressor first acts as a pump, circulating refrigerant through the system, and secondly as a compressor, compressing the refrigerant so that it can be more easily changed from a gas to a liquid. Several different types of compressors will be used by various manufacturers, but all will have the same general appearance and the same general characteristics. Compressors used in residential air conditioning will range from 1½ tons to 5 tons of refrigeration, Figure 9-1. They are hermetically sealed in a steel shell and, if there is an internal malfunction, they are *not* serviced or repaired in the field. If the compressor has failed for any reason, it is removed from the system and, if in warranty, returned to the facto-

FIG. 9-1.   Typical hermetic compressors.

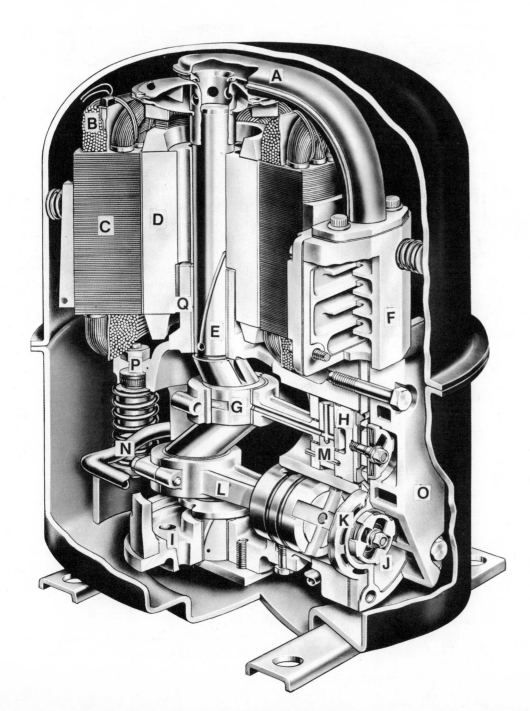

FIG. 9-2. Cutaway view of hermetic compressor. *Courtesy Tecumseh Products Co.*

A. The suction tube to the compressor cylinder is attached over an oil slinger or separator.

B. Internal thermostat, imbedded in the motor windings, has two leads to the electrical terminal block on the compressor. The internal thermostat senses motor heat only and is wired into the control circuit.

C. Motor stator containing both the start and run windings.

D. Motor rotor, which is press-fit onto the compressor crankshaft.

E. Crankshaft is offset to provide an eccentric or throw for the pistons.

F. Suction muffler has a series of baffles through which the suction gas must pass. In so doing, its noise level is reduced.

G. Connecting rod main bearing.

H. Cutaway view of the piston. Note that there are no rings on the piston. The grooves hold a film of oil to seal the cylinder.

I. Bottom or cage bearing.

J. Discharge valve.

K. Suction valve into the cylinder.

L. Connecting rod with bolted cap.

M. Wrist pin joins the piston to the connecting rod. Note that this pin is drilled to provide lubrication to the cylinder walls.

N. Discharge tube from the cylinder head.

O. Cylinder head.

P. One of three internal mounting springs.

Q. Top bearing.

ry for repair and credit. Therefore, it is not necessary for the serviceman to completely understand the internal parts of a compressor. However, it is important to have an idea of how a compressor works, in order to recognize problems. Compressors are remarkably free of any mechanical defects and, if the original installation is made carefully and properly, they will last for quite a number of years without failure. Most manufacturers experience an extremely low failure rate on the compressors. The majority of apparent failures usually can be attributed to a bad installation. A very high percentage of the compressors that are returned under warranty have nothing wrong with them internally.

The internal parts of a basic hermetic compressor are shown in Figure 9-2. This particular compressor is a Tecumseh CL compressor which is built in the 3 to 5 ton range.

LUBRICATION

This particular compressor will have a total oil charge of 55 ounces. Some smaller air conditioning compressors will have an oil charge of about 45 ounces. The oil level, with the compressor not running, will be approximately to the bottom of the lower cylinder, Figure 9-3. On start-up, the oil will be changed into a fine mist and mixed with refrigerant which is then carried through the entire system. For this reason it is important that the lines be sized and bent so as not to restrict the flow of oil returning to the compressor crankcase. The slot and hole at the bottom of the crankshaft is the oil access opening to the

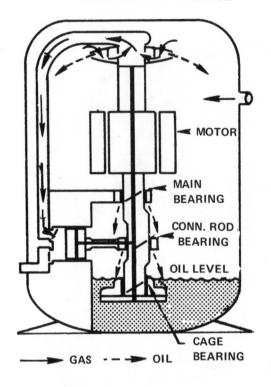

FIG. 9-3. Oil and gas flow paths.

lower bearing. Another hole is drilled in the bottom of the crankshaft to carry the oil up to the connecting rod where it exits at *Point G*, Figure 9-2. There is a groove milled in the center of the connecting rod that carries oil to the center of the rod shaft and then out to the wrist pin at *Point M*. Here the oil is carried to the compressor cylinder, lubricating this area. Oil from the center of the crankshaft also exits at the top main bearing, approximately *Point E*, and is carried by a slot

that lubricates the entire bearing surface. The oil going to the bearings will drip down and return to the oil reservoir at the bottom. The oil going to the pistons and cylinder must go through the entire system before returning to the crankcase.

REFRIGERANT GAS FLOW

Suction gas enters the compressor shell, usually at the top, through the suction tube. The suction tube opens directly into the compressor shell and there is no additional tubing. Thus, the compressor shell is subjected to the low side pressures of the system which, for R-22, will run between 75 and 85 psig. Suction gas totally fills the inside of the shell and removes some of the heat of the motor. It is drawn into the cylinder through the oil separator at *Point A*, Figure 9-2.

The oil separator, or anti-slugging device, takes in gas and oil mixtures through the slots on top and, as it rotates, it forces the heavier oil out of the edge slots by centrifugal force, Figure 9-4. Thus, the majority of the oil is returned to the crankcase rather than going through the cylinder.

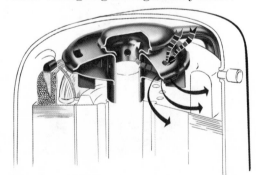

FIG. 9-4. Oil separator or anti-slugging device. *Courtesy Tecumseh Products Co.*

Gas leaving the separator follows the tube down into the muffler, *Point F*, Figure 9-2. It enters the compressor cylinder through the suction valve, *Point K*, which, of the two valves in the cutaway, is closest to the piston. There it is compressed by the piston. The compressed gas exits through the discharge valve, *Point J*, going into the discharge tube which is designed for high pressures. The discharge tube is coiled once, to create a vibration loop, then is connected to the shell at the external discharge line. Most compressors will take in suction gas near the top third of the shell and discharge it near the lower third.

Externally, the compressor will have one discharge and one suction tube. There may be a third tube, which is a process tube, that has been used at the factory for initial processing. This tube is not used for any field connections and if it is plugged, as it frequently is on a replacement compressor, it must be pinched off and silver soldered. The compressor will be mounted on 3 or 4 rubber grommets on its external feet, and will have a terminal block for electrical connections which will be discussed later.

FLOOD BACK

Under certain operating conditions, liquid refrigeration may enter the compressor through the suction tube. This liquid will fall to the bottom of the compressor and mix with the oil, diluting it. If the oil is diluted to any great degree, its lubricating ability is reduced and this can cause bearings to overheat. If the heat becomes excessive, approximately 300°,

then the refrigerant and oil mixture can break down into hydrochloric acid. This acid will have a corrosive effect on the motor windings, potentially causing a motor burnout. Here are some possible causes of liquid flood back:

1. Overcharge of refrigerant.
2. Undercharge of refrigerant (flood back occurs after the coil frosts up).
3. Restriction of evaporator air.
4. Very low indoor temperature or relative humidity, reducing the load.
5. Defective expansion valves.
6. Very low outdoor temperature.

Where these conditions exist, it may be necessary to add a suction line accumulator to the system. This is a small tank inserted into the suction line immediately ahead of the compressor, Figure 9-5. The accumulator acts as a trap for liquid refrigerant, which drops to the bottom of the tank, but the gas can still be drawn into the compressor through the outlet tube. A small hole at the bottom of the tube allows oil to be picked up by the vapor refrigerant as it passes through, so that it will be returned to the compressor crankcase.

MIGRATION

When a compressor is not operating, particularly in winter, most of the refrigerant condenses into a liquid. This liquid will migrate to the coldest part of the system. This is normally the compressor crankcase, since the crankcase is outside and the bottom of the compressor shell is usually on a slab or close to the ground. The oil, of course, is in the compressor crankcase and, as it gets colder because of

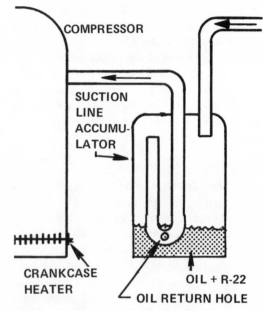

FIG. 9-5. Accumulator stops flow of liquid refrigerant into the compressor.

ambient temperature, it attracts more and more refrigerant which condenses in the compressor shell. Over a period of time, most of the refrigerant in the system will be attracted to the compressor and condense into liquid.

When the compressor starts up, with the shell filled with liquid, the rapid reduction in pressure causes violent boiling of the liquid refrigerant-oil mixture. Even with an anti-slug device on the compressor, this refrigerant-oil mixture can be drawn into the cylinder. Because it is a noncompressible liquid, it can cause *slugging* of the compressor or blown gaskets. Even if slugging does not occur, a great deal, if not all, of the oil will be carried through the entire system, forcing the compressor to operate for several minutes with an insufficient oil supply. This lack of lubrication can be damaging to the bearing surfaces of the compressor.

This situation can be prevented by several methods, all of which add heat to the oil in the compressor crankcase during the period when the compressor is shut down.

CRANKCASE HEATERS

There are three basic methods of adding heat to the compressor crankcase to prevent refrigerant migration:

1. A crankcase heater is an electric heater element of about 50 watts, which is wrapped around the base of the compressor and wired so that it always has electrical current, even when the compressor is off. This heater can hold the inside temperature at about 10° above the ambient, which is enough to prevent the refrigerant from condensing in the crankcase, Figure 9-6.

   A similar method is a blanket type heater. This is also a resistance heater, but has a greater surface area, and is placed beneath the compressor crankcase, rather than wrapped around it.

2. An oil rectifier is an electric resistance heater, in the form of a rod element, that is inserted into the compressor crankcase near the lower third of the normal oil level. The compressor will have a well to accept the crankcase heater. The rod is inserted into this well and wired directly to the line side of the contactor so that it will be operative whenever the main disconnect switch is on, Figure 9-7.

FIG. 9-6.   Crankcase heater is wrapped around the base of the compressor. *Courtesy The Coleman Co., Inc.*

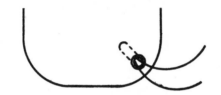

FIG. 9-7.   Rod-type crankcase heater is inserted into a well in the compressor housing.

3. Run capacitors can be wired so that they impose a low voltage on the compressor motor windings during the off cycle. A run capacitor, which has a known and constant impedance, is located in one leg of the line. This causes a very large voltage drop across the capacitor terminals and a small voltage drop across the compressor terminals, allowing a small current to flow through the motor windings. Since the motor is immobile, this current is dissipated as heat.

Capacitors are matched to the motors so that they will create the required temperature differential, due to the electrical input to the winding, between the compressor and the surrounding ambient. This temperature difference will maintain a higher crankcase temperature and prevent migration of any sizable quantity of refrigerant to the compressor housing.

Some systems split the total run capacitance between two capacitors. Power to the motor windings is supplied through one of them. With the overload located in the common side of the line, a single pole contactor may be used to be sure that power is always applied. Because it will have a continuous voltage, a fused capacitor must be used for maximum compressor protection. These are available either in a single can, internally fused, or as a two section capacitor with three terminals, Figure 9-8.

FIG. 9-8. Two section run capacitor.

NOTE: Power must be supplied to the condensing unit at all times in order for the crankcase heating device to be operative. Most installations will carry a sticker which is put on the service disconnect at the condensing unit. It states that the main power should be in the on position for at least 24 hours before operating the compressor.

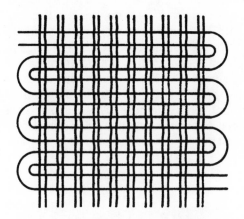

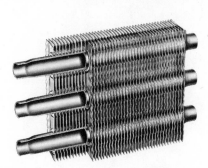

FIG. 9-9.   Condenser consists of copper tubing arranged in a serpentine pattern. (Below) Cutaway of condenser tubes with fins.

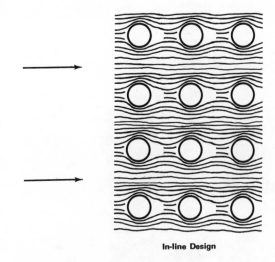

In-line Design

FIG. 9-10.   Condenser tubes are usually staggered (below) to increase heat transfer.

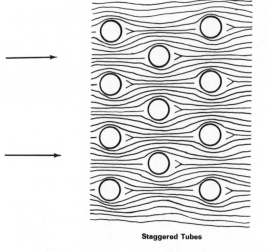

Staggered Tubes

CONDENSERS

The function of the condenser is to remove the heat from the refrigerant by passing cooler outside air over the heat transfer surface. Refrigerant will change from a gas to a liquid as it passes through the condenser and, at the exit, will be 100% liquid. This is piped through the liquid line to be ready for another trip through the system.

The condenser consists of copper tubing, arranged in a serpentine fashion and 2 or 3 rows deep, Figure 9-9, which is the primary heat transfer surface. Attached to the tubing will be thin aluminum fins, spaced anywhere from 6 to 14 fins per inch of tube. These act as a secondary surface to increase the effectiveness of the heat dissipation. Several rows of tubes and fins are common. The tubes will usually be staggered to get the maximum air turbulence and heat transfer from the passing air, Figure 9-10. Outside or ambient air is forced over the surface of the condenser by a condenser fan which, in residential sizes, will move from 1000 to 2000 cfm of air over the fins. Condensers can be mounted horizontally, with the air blown from bottom to top, or they may be arranged vertically, with the air being either blown through or sucked through.

The condenser and the condenser fan are usually contained in a single housing which also contains the compressor, the service valves, the drier and electrical components. This whole assembly is called the condensing unit, the high side, or the outdoor unit, Figure 9-11.

EVAPORATORS

An evaporator has two basic functions: one is to reduce the temperature of the indoor air passing over it, the second is to reduce the moisture content and dehumidify the air. In doing this, the refrigerant changes from a liquid to a gas while passing through the evaporator. The evaporator resembles the condenser in its construction and may be 2, 3 or 4 rows deep.

Condensate is formed as part of the dehumidifying process. As the air is cooled, its ability to hold water vapor is reduced. Some of this airborne water vapor condenses out as water which must be piped to a drain connection, usually through a plastic hose. The evaporator is placed in the furnace plenum, immediately in the supply airstream going to the house, Figure 9-12.

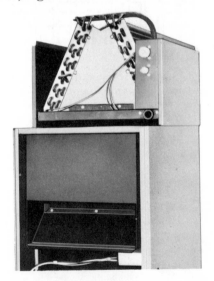

FIG. 9-12. Evaporator is placed in furnace plenum. *Courtesy Bard Mfg. Co.*

FIG. 9-11. Basic condensing unit components. *Courtesy The Coleman Co., Inc.*

1. Vertical air discharge.
2. Outside air drawn through condenser.
3. Multiblade fan.
4. Insulated compartment to reduce noise.
5. Liquid line service valve.
6. Compressor.
7. Resilient rubber grommets under each leg.
8. Unit support channels.
9. Fan guard.
10. Weatherproof steel cabinet.
11. Grill in front of condenser fins.
12. Low pressure control.
13. Contractor, compressor run capacitor, fan motor capacitor.
14. Liquid line filter drier.
15. Compressor terminal cover.
16. Crankcase heater.

There are two common types of eva-porators in residential systems. One is an *A coil*, which has two evaporators connect-ed together and mounted at an angle to each other, roughly forming the letter *A*. Hence, its name. An A coil is primarily used on an up-flo furnace but it also can be used with a horizontal or down-flo fur-nace. The second type is a *slab coil* which is built as a flat rectangle. It can be mounted vertically, within horizontal type furnaces, or the slab coil can be mounted at an angle for up-flo, down-flo or horizon-tal applications, Figure 9-13.

Courtesy Addison Products Co.

Courtesy The Coleman Co., Inc.

Courtesy Addision Products Co.

FIG. 9-13.   Types of condensers. (Above) A coil. (Top right) Slab coil. (Lower right) Slant coil.

CONNECTING TUBING

The evaporator, condenser and com-pressor are connected with a series of cop-per tubes to form a complete refrigerant circuit, Figure 9-14. The *suction line* connects the evaporator to the compres-sor. Since this line must carry gas, which has a greater volume than liquid, it must have a larger diameter to compensate for this. This line will be insulated because it will be colder than the ambient air and will have a tendency to collect condensate (water) on its surface.

A second line will run from the com-pressor discharge valve to the inlet side of the condenser. Since the compressor and condenser are mounted in the same cabi-net, this will be a relatively short line. It is called the *hot gas* or *discharge line*. It will be smaller than the suction line because the gas is now compressed into a smaller volume. It will be much hotter than the other lines in the system because the gas

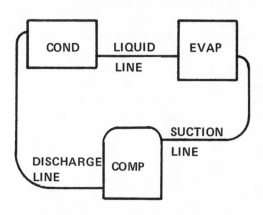

FIG. 9-14.   Connecting lines used to link the major system components.

has its normal heat content, additional heat from the compressor motor, plus the heat of compression. Heat of compression is what causes an automobile tire to get hot when it is inflated (increasing its internal pressure).

A third line will run from the outlet of the condenser, which is usually at its lowest point, back to the inlet side of the evaporator. This line carries liquid refrigerant, which has been condensed in the condenser. It is called the *liquid line.*

This closed refrigeration system has one major purpose, to recycle and reuse the refrigerant in the system. It is possible to obtain air conditioning by merely using a coil or some other heat exchanger. The refrigerant would pass through the coil, picking up heat, and could be expelled to the outside air. While this system would work, it would be wasteful since the refrigerant would only be used once. And, the coil would freeze because R-22 boils at -41° F at atmospheric pressure.

Additional components are placed in the system so that the refrigerant can be recovered and reused over and over again. Closing the system allows the pressure in the evaporator to be held to about 77 lbs per sq in. gauge (psig). At 77 psig, the refrigerant will not freeze the coil since the corresponding temperature is 45° F.

DRIERS

Even though extreme care is taken during manufacture and installation, almost all manufacturers install a liquid line filter-drier, Figure 9-15, in an air conditioning system. A filter-drier is designed to remove both solids and undesireable

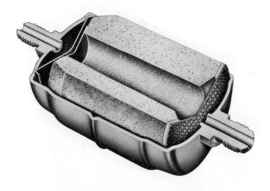

FIG. 9-15. (Above) Cutaway of filter-drier. (Below) Filter-drier installed in liquid line. *Courtesy Sporlan Valve Co.*

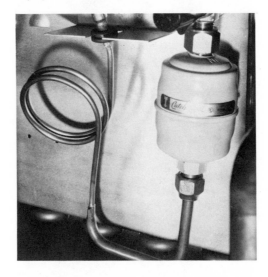

solubles, like water and acid, from the refrigerant. There are several types of filter-driers but most combine an activated alumina dessicant (drying agent) with a molecular sleeve. A metal screen will be used at the entrance, to trap solids, and

the design of the core assures that all of the refrigerant will come in contact with the drying agents. Filters are mounted in the liquid line, usually near the condensing unit. Whenever the compressor must be changed because of a burnout, the liquid line drier should also be changed. It is also good practice to change the drier on an older system which has been opened or is suspected of containing moisture.

PRESSURE CONTROLS

Most compressors will include a low pressure control and some also will have a high pressure control, usually mounted on the compressor shell. A small tube or capillary will be connected to a port on the service valve so it can sense pressure under operating conditions. Both controls may be combined into one control called a dual pressure control. They are wired electrically in series into the low voltage control circuit.

The low voltage control is in the circuit primarily to protect against loss of refrigerant charge. However, if the suction pressure goes below the control setting (about 5 to 8 psig) for any reason, it will trip, shutting off the compressor. It is an automatic reset control. That is, it will reset itself when the suction pressure raises above the setting.

The high pressure control is in the circuit to protect against failure of the condenser fan motor or any other condition which causes the head pressure to increase to an unsafe point. It usually has a manual reset feature which means that once it trips, it must be reset manually before the compressor will run.

METERING DEVICES

The control or metering device must match the flow of refrigerant to the load on the evaporator, the pumping capacity of the compressor and the ability of the condenser to reject heat. It determines the capacity of the system which is probably the most important function in the entire system.

While a compressor is a constant displacement pump, that is, it has a fixed cylinder capacity plus a constant, 3500 rpm operating speed, the volume of refrigerant pumped can vary with the density of the refrigerant in pounds per cubic foot. Since the refrigerant, as it goes through the compressor, is in the vapor state, its volume and density will be affected by its temperature. As temperature increases, the volume increases and therefore the pumping rate of the compressor, in terms of lbs/hr pumped, will decrease due to decreased density. The temperature of the refrigerant gas as it reaches the compressor is dependent upon the load upon the evaporator, since the greater the load, the more the refrigerant temperature will increase.

Each time a pound of refrigerant makes the circuit through the refrigeration system, the system will try to remove as much heat as possible. There are two basic factors which will affect the ability of the system to remove heat and thus increase or decrease capacity and flow rate of the refrigerant. These are:
1. Changes in the temperature or quantity of outdoor air.
2. Changes in the temperature or quantity of indoor air.

OUTDOOR AIR

The quantity of outdoor air is fixed due to the constant speed of the condenser fan. Therefore it is only necessary to consider changes in temperature of the outside air. As the outside air temperature becomes higher, it accepts less heat from the refrigerant because the temperature difference between the refrigerant and the outside air is smaller. Therefore, the compressor must pump the gas to a higher temperature (head pressure) to get it into the condenser. This reduces the quantity of refrigerant the compressor can pump and the metering device therefore must allow *less* refrigerant to flow in order to equal the lower pumping rate. This situation reduces the overall capacity of the system.

As the outside air becomes colder, the reverse happens and the compressor is able to pump more refrigerant. This is because the ability of the condenser to reject heat is now greatly improved and therefore the metering device must allow more refrigerant to flow. In this instance the capacity of the system will be increased.

INDOOR AIR

In a similar manner, the quantity of indoor air passed over the coil will remain reasonably constant, since the speed of the blower is constant. However, as the indoor air becomes warmer, it is lighter in weight and therefore the same blower speed can pass a slightly greater volume of air over the coil. Since the air has more heat content, it is easier for the coil to extract this heat from the air and thus the metering device must allow a greater flow of refrigerant into the coil. If the indoor air becomes colder, it also becomes heavier so the quantity or volume of air over the evaporator will be slightly less. It is not possible to extract as much heat from the air when it is colder, because the difference in temperature is smaller. Therefore, the metering device must reduce the amount of refrigerant flowing through the system.

The metering device must always reduce the flow of refrigerant if the outside air temperature increases or if the indoor air temperature decreases. Conversely, it must increase the flow of refrigerant if the outside air temperature decreases or if the indoor air temperature increases. A change in outdoor air temperature varies with weather conditions and a change in indoor air temperature varies with the load inside the house. This load can be infiltration, which increases as outside air temperature goes up, people, lights, cooking or anything else that increases the indoor air temperature.

All of this assumes that the unit is properly sized for the outside design conditions and the internal load. Proper sizing will take care of some variations in these factors but, if the unit is undersized for the load, it will not be able to hold temperatures at the level set by the homeowner. Units are normally slightly undersized so that they will run longer and so remove more moisture but, if too much undersize, they provide inadequate temperature control.

There are two metering devices used on residential air conditioning systems.

Each adjusts the refrigerant flow rate, but in quite a different manner. The first is the expansion valve.

EXPANSION VALVE

An expansion valve is a metering device used on many air conditioning systems, although it is normally not used with under 4-ton systems, Figure 9-16. The expansion valve controls the flow of refrigerant by means of a needle valve placed in the refrigerant line.

Liquid refrigerant enters the valve from the high pressure side of the system and passes through the needle valve to the low pressure side. Here its pressure is reduced, causing a portion of the refrigerant to vaporize immediately, cooling the balance of the refrigerant at this point. The refrigerant passes through the evaporator, absorbing heat from the indoor air passing around the evaporator, and by the time it reaches the exit it will be almost entirely gas.

Some expansion valves will have a series of tubes at the exit so that they may feed more than one row of an evaporator. These are called distributors and are designed to feed the several rows of a multi-circuited evaporator simultaneously, Figure 9-17.

The thermostatic expansion valve has three major functions:

1. *Throttling Action.* A thermostatic expansion valve separates the high and low sides of the system. This pressure difference, between the condenser and evaporator, is maintained by the valve to assure the most efficient system performance.

FIG. 9-16.   Thermostatic expansion valve. *Courtesy Sporlan Valve Co.*

FIG. 9-17.   Distributor allows expansion valve to feed several rows of the evaporator simultaneously. *Courtesy Sporlan Valve Co.*

2. *Modulating Action.* The valve feeds liquid refrigerant into the evaporator, in the proper amounts, at all times. If too much liquid refrigerant enters the evaporator, not all of it will change to a vapor and there is danger of passing liquid into the compressor, damaging the compressor valves. If there is too little refrigerant in the evaporator, all of the liquid will be evaporated before it makes a complete circuit, reducing the system capacity and *starving* the evaporator. By modulating the flow, the valve will maintain the proper amount of liquid in the system.

3. *Controlling Action.* The valve must also respond to load changes in the system. As the load in the system increases, the valve moves to the wide-open position, lowering the pressure drop across the valve port and allowing more refrigerant to flow. Thus, the compressor capacity balances with the increased load on the system. As the load decreases, the valve closes, reducing the amount of refrigerant available to the compressor.

The opening and closing of this valve is controlled by a combination of forces that

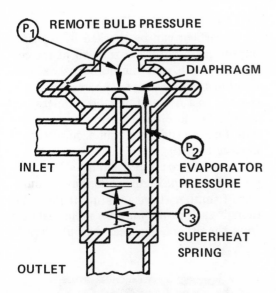

FIG. 9-18. Three pressures are used to balance expansion valve flow rate with evaporator demand.

constantly monitor the temperature of the refrigeration system. An expansion valve consists of an outlet and inlet for the refrigerant and a diaphragm separating the evaporator inlet and outlet pressures.

In addition, there is a thermal bulb which is attached to the suction line at the exit of the evaporator. This bulb is filled with the same refrigerant as is used in the system. The bulb senses the temperature at the evaporator exit and is connected to the diaphragm by a capillary tube. Therefore, the operation of the valve is affected by three basic forces, Figure 9-18:

1. Bulb temperature (pressure at the exit of the evaporator) acting on one side of the diaphragm, which tends to open the valve (P_1).

2. Evaporator or suction pressure acting on the opposite side of the diaphragm, which tends to close the valve (P_2).

3. Spring pressure on the evaporator side of the diaphragm also tends to close the valve (P_3).

When the pressures above and below the diaphragm are equal, the valve is said to be in equilibrium. That is, it will pass the correct amount of refrigerant required by the conditions the system. The remote bulb senses the leaving temperature of the refrigerant gas as it passes out of the evaporator. This gas should contain no liquid which, being noncompressible, could damage the compressor.

Therefore, the pressure exerted by the sensing bulb is offset by the adjustable spring at a pressure that assures that the leaving vapor is superheated and contains no liquid. The amount of superheat is determined by consulting a Saturated Temperature/Pressure Chart. Superheat is the number of degrees by which the unsaturated vapor (containing no liquid) exceeds the temperature of a saturated vapor (containing liquid) at the same pressure. In residential air conditioning systems, the spring is set to assure from 6° to 15° of superheat, with an average of about 10°.

As system load changes, the suction temperature and pressure of the inlet and outlet of the evaporator will change but the *difference* in temperature or superheat should remain the same.

Multicircuited evaporators will also have an accumulator or manifold to collect the refrigerant gas from the several circuits and pass it into the single suction line, returning to the compressor. In this instance it is important that the feeler bulb will be strapped to the manifold or final collecting point and not one of the

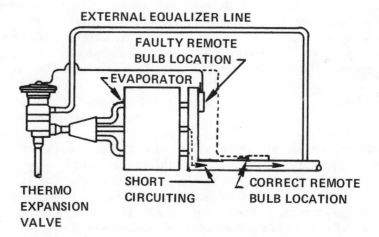

FIG. 9-19. Correct locations for the remote bulb and the equalizer line.

individual circuits, Figure 9-19.

If there are no unusual pressure drops in the evaporator (normal design pressure drop is 2 psig), the inlet gas pressure is reasonably close to the exit gas pressure. In this case, the valve senses pressure at the evaporator inlet, Point P_2 in Figure 9-18.

If the pressure across the evaporator exceeds 2 lbs, an excessive amount of inlet gas pressure offsets the pressure generated at the remote sensing bulb. In one example, 6 lbs pressure drop required an additional 6° of superheat to offset the pressure on the diaphragm. While the system was attempting to achieve equilibrium at this higher superheat, the evaporator was *starved* for lack of sufficient refrigerant.

To correct this problem, an external equalizer is tapped into the suction line, at a point downstream from the sensing bulb, and connected to the underside of the valve diaphragm, Figure 9-19. In this way, the effective pressure at the bulb is fed to the valve and superheat is maintained at a normal level. As the result, the evaporator receives the required amount of refrigerant, despite the pressure drop through the coils.

INCREASE IN LOAD

If the load on the evaporator increases, the refrigerant passing through the coil will pick up more heat and will be at a higher temperature than the refrigerant at the inlet of the evaporator. The expansion valve bulb at the exit of the evaporator will sense this and exert additional pressure on the diaphragm, opening the valve and allowing more refrigerant to flow. Liquid refrigerant at the entrance to the evaporator will evaporate at a faster rate, raising the evaporator pressure and bringing the forces again in balance at the new setting. Thus, the evaporator pressure acting on one side of the diaphragm and the additional bulb pressure acting on the other will tend to offset each other, and the valve will adjust to the new load conditions. The additional pressures on both sides will tend to open the valve, since more pressure is exerted in that direction even though the superheat setting or difference in temperature remains the same.

DECREASE IN LOAD

If there is a very light load on the evaporator, the liquid admitted to the evaporator will not completely vaporize until it almost reaches the outlet of the coil. Therefore, its temperature and pressure will be very close to the temperature and pressure at which it entered. This reduction in temperature will be sensed by the feeler bulb, which will exert less pressure on the diaphragm. The valve will tend to close, reducing the flow of liquid. Although the inlet and outlet temperatures and pressures are lower, the difference in pressure will remain the same as the superheat setting, so the valve can adjust to this change in load.

In all cases, the valve adjusts to changes in load by changes in refrigerant flow. This flow rate then matches the pumping rate of the compressor so that the system is always in balance and maintains its basic capacity.

CHECKING SUPERHEAT

Superheat can be checked by measuring the temperature at the exit of the coil near the bulb, and simultaneously checking it at the entrance to the evaporator. If the coil has a manifold, the temperature should be taken at the line beyond the manifold in order to get an accurate reading, Figure 9-19.

Test instruments, such as the Amprobe Fastemp, Figure 9-20, have a number of temperature sensing probes. Although each probe may be some distance from the others, each can be quickly and

FIG. 9-20.   Amprobe Fastemp has three temperature sensing probes which can be read in split seconds. *Courtesy Amprobe Instrument.*

conveniently read by switching to the desired probe and reading the temperature shown on the dial. Here are some typical readings:

Evaporator Exit Temperature	50°
Evaporator Inlet Temperature	40°
Superheat	10°

Where this type of temperature recorder is not available, the superheat can be determined as follows:

1. Strap a thermometer on the suction line near the bulb from the expansion valve. Be sure good contact is made and pack some Permagum or other substance around the thermometer bulb to insulate it from ambient air, Figure 9-21.
2. Attach a gauge set to the unit (method is described in Chapter 10).
3. After the unit has run long enough to stabilize (5 to 10 minutes) read the suction pressure and convert to temperature using Temperature/Pressure chart (Table 8-1).
4. Read the temperature at the exit from the evaporator.

For example:

Evaporator Exit Temperature	53°
Suction Pressure 70 psig	
Saturated pressure equals	41°
Superheat	12°

Modern expansion valves are quite dependable and manufacturers caution

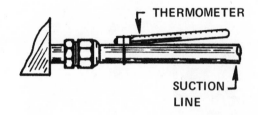

FIG. 9-21. Thermometer is attached to suction line and packed with Permagum before measuring superheat.

that other factors in the system which can affect superheat be checked before concluding that the valve is bad or the superheat setting should be changed. For example, if a system is low on charge, one symptom will be a higher than normal superheat reading. Also, be sure to check the manufacturer's recommended superheat setting, since they can vary over a wide range depending upon the system and type of valve used.

CAPILLARY TUBE SYSTEMS

The second basic type of metering device is a capillary tube, which is nothing more or less than a length of copper tubing with a very small inside diameter.

Capillary tubes are commonly used on small size (3 tons and below) refrigeration systems of all types. It is the standard control device for household refrigerators, freezers, etc. In recent years, its use has been expanded to residential air conditioning and it has been successfully used with up to 5-ton systems.

The capillary tube may be built into the evaporator or it may be attached to the end of the liquid line. When connected, it links the liquid line with the evaporator.

The length and diameter of this capillary tube are very carefully matched to a specific unit so that it will precisely meter the refrigerant required by the system to which it is attached.

The length of this capillary tube is critical and it cannot be shortened in the field. Even if the tube is 6 to 8 feet long, as it often is with a single capillary system, it is coiled up neatly, and left intact. The internal diameter of these restrictors, or capillary tubes, is quite small. In the residential air conditioning sizes, several will be used in the neighborhood of .080 to .090 ID. Multicircuited evaporators will have a capillary tube for each circuit, Figure 9-22. These will each be 1 to 2 feet long and in the same size range as single tube systems.

Capillary tubes have come into widespread use for several reasons. One is basic cost. Another is that they will equalize pressures in the system on the off cycle. Therefore, the compressor motor does not have to start against a full load. This allows reduction in motor torque and also starting components, which again enters into the total cost of the system.

At first glance, it would appear that a small diameter tube, with no valves or other restrictions to control the flow, would deliver refrigerant at a constant rate. However, this supposition overlooks the fact that increasing amounts of vapor are produced as the temperature of the refrigerant is lowered. For instance, saturated R-22, at 45°, contains twice as much vapor, by volume, as R-22 at 88°.

While both liquid and vapor are delivered, the performance of the evaporator

FIG. 9-22.   Multicircuited evaporators have a capillary tube for each circuit. *Courtesy Friedrich Refrigerators, Inc.*

depends on the amount of liquid available, since vapor has practically no refrigerant effect. Therefore, the performance of the evaporator is controlled by delivering more vapor and less liquid when the evaporator has attained its design temperature and more liquid and less vapor when the evaporator requires refrigerant.

An irritating example of this principle is *vapor lock* which occurs when the fuel delivery system of a car is overheated. Fuel flow does not cease when there is vapor lock. Rather, the carburetor, which demands liquid, is receiving vapor rather than liquid. If it receives a mixture of liquid and vapor, the engine starts missing, Figure 9-23. If it receives only vapor, the engine will stop.

FIG. 9-23.   Flow rate is reduced when vapor displaces liquid in a liquid feed line.

As a practical example, assume a condensing temperature of 130° F, with a saturated pressure of 300 psig, and an evaporating temperature of 45° F, with a saturated pressure of 76 psig. Incidentally, these points are used by the ARI (Air Conditioning and Refrigeration Institute) for comparative ratings of all residential air conditioning systems.

First, the pressure must be reduced from 300 to 76 psig, a pressure drop of 224 psig. However, if we were to reduce the inlet pressure to 250 psig, the outlet pressure, at 45°, would still be 76 psig. So the performance of the capillary tube is not necessarily dependent on pressure drop across the tube.

Temperature, therefore, is our principle concern. The temperature drop is from 130 to 45°, a total of 85°. To effect this temperature drop, liquid must boil in the capillary tube. In so doing, it refrigerates the liquid and reduces its temperature. The total amount of vapor produced throughout the tube is, in a large sense, related to the temperature drop within the capillary tube. A greater temperature difference between the condenser and the evaporator produces more vapor in the capillary tube. Reducing the temperature difference between the condenser and the evaporator, reduces the percent of vapor in the capillary tube.

When the indoor temperature equals the design temperature of the evaporator, which is 45°, the percent of vapor is at a maximum. The temperature drop is also at a maximum. However, with a sudden increase in indoor temperature, the evaporator temperature could rise to 55°. As it does, the temperature of the refrigerant increases, the percent of vapor decreases and the temperature difference is reduced by 10° . As a result, more liquid refrigerant will flow into the evaporator and lower its temperature.

Further increases in room and evaporator temperature will increase the percentage of liquid in the liquid-vapor mixture and, therefore, its refrigerating effect. The percentage of liquid will again decline as the evaporator matches the requirements of the heat load and begins to return to its design temperature of 45° .

You can actually feel these changes in the temperature of the capillary tube. If you place your hand on the tube, where it enters the evaporator, you can feel the tube become warmer as the load on the evaporator increases. It will become cooler as the load decreases.

If you run your hand along the line, moving back towards the condenser, the capillary will become progressively warmer. However, there is a point where there is no further temperature increase. This is called the *bubble point.* Between this point and the evaporator, the tube is filled with the liquid-vapor mixture at saturated temperatures and pressures. Between this point and the condenser, the tube is filled with liquid, containing no vapor. This liquid is subcooled since the pressure exceeds the temperature at which a saturated condition can occur. Incidentally, this subcooling is the result of heat transfer between the refrigerant in the tube and the ambient air.

The location of this bubble point is important to the flow rate of the refrigerant since the liquid flows at a far faster rate than the liquid-vapor mixture. If the distance from the condenser to the bubble point has increased, indicating that the condenser temperature has decreased, the rate of flow increases, while the amount of vapor in the line between the bubble point and the evaporator decreases. If the distance between the bubble point and the condenser decreases, indicating that condenser temperature has increased, the rate of flow decreases, since the amount of vapor in the line between the bubble point and the evaporator has increased.

This shift in the bubble point indicates that the capillary tube is adjusting to the demands of compressor capacity. For instance, if the outdoor temperature increases, the heat-rejecting capabilities of the condenser may be exceeded. At the same time, compressor capacity may be exceeded. Therefore, the supply of refrigerant to the evaporator must be reduced. And it is, automatically, because the bubble point moves closer to the condenser, reducing the length of the liquid line, increasing the length of the two-phase line, which contains both liquid and vapor, and slowing the flow of refrigerant. As a result, evaporator temperature will slowly rise, even though the pressure difference between the inlet and outlet of the capillary has actually been increased.

If the outdoor temperature decreases, the condenser can handle greater quantities of refrigerant and the demand is once again within the capacity of the compressor. The flow of refrigerant can now be increased. And it is, automatically, because the bubble point moves toward the evaporator, increasing the length of the liquid line, decreasing the length of the two-phase line and increasing the flow of refrigerant.

The bubble point is indicated on the Pressure-Enthalpy diagram previously shown as Figure 8-7. It occurs at the point where the pressure-enthalpy line, A-B, crosses the saturation line.

As the line A-B moves to the left, the liquid portion of the line becomes longer and flow increases. As line A-B moves to the right, the liquid portion of the line becomes shorter and flow is reduced.

A capillary tube system is usually very critical on charge and, as was pointed out before, very critical on the length and diameter of the capillary tube itself.

Because of its very small diameter, cleanliness in the system is of major importance since a very small particle of dirt can plug up the capillary tube, causing ti to malfunction. Most capillary tube systems are factory-assembled in order to control these problems, but field-assembled systems should be installed very carefully to make sure that no dirt or moisture enter the system.

Questions

1. What is done with a compressor that fails within warranty?
2. What is the purpose of an oil separator?
3. What is the purpose of an internal thermostat?
4. Will some oil be carried through the system? Why?
5. Where is the internal suction tube?
6. What is an important function of the suction gas inside of the compressor shell?
7. Name some possible causes of liquid refrigerant entering the compressor.
8. How can this situation be remedied?
9. What is meant by migration?
10. How can this become a problem?
11. What is the solution to refrigerant migration?
12. What is included in the condensing unit?
13. What is the function of the condenser?
14. What are the two functions of the evaporator?
15. Name the connecting lines for a refrigeration system.
16. When is a drier installed in the system?
17. What are the two most common types of metering devices?
18. What are the two basic areas which affect the ability of the system to increase or decrease capacity?
19. If outside air temperature increases, what must the metering device do and why?
20. If the outside air temperature decreases, what must the metering device do and why?
21. If the indoor air temperature increases, what must the metering device do and why?
22. If the indoor air temperature decreases, what must the metering device do and why?
23. What are the three major functions of the expansion valve?
24. What is the function of the distributor?
25. Give the three basic forces controlling the operation of an expansion valve.
26. What is the average superheat setting in a residential system?
27. What is the purpose of an external equalizer?
28. How can superheat be measured?
29. What is a capillary tube?
30. What are two major reasons for using capillary tubes?
31. What is the principle used for capillary tube operation?

Chapter 10
COOLING INSTALLATIONS

Two terms used in the industry are *high side* and *low side*. These terms refer to the pressures normally found in the two halves of the air conditioning system and have nothing to do with whether one section of the system is physically higher or lower than the other.

The low (pressure) side of the system runs from the exit of the metering device, through the evaporator and into the compressor shell. The high (pressure) side of the system begins at the compressor cylinder, and runs through the discharge tube to the condenser and back through the liquid line to the entrance of the metering device.

The term *split system* or *remote system* is used when the high side components are mounted in their own cabinet and placed outside the conditioned space. In this case, the low side is mounted in the furnace and the furnace blower becomes the prime air mover, Figure 10-1.

In a self-contained system, the condensing unit, plus the evaporator, the condenser fan and evaporator blower, are all mounted together in a common cabinet, Figure 10-2. This unit is independent of the heat source, although it can operate

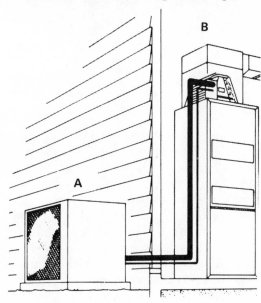

FIG. 10-1. Typical split system. (A) High side components. (B) Low side.

through the same duct system. More often than not, it is applied where the heating equipment does not have the capability of handling the amount of air necessary for cooling or where there is no forced air heating system.

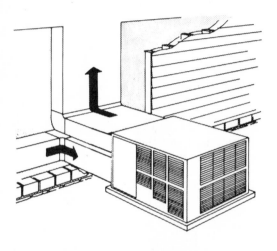

FIG. 10-2. Typical self-contained system has all components, plus the evaporator blower, mounted in a single cabinet.

Some self-contained systems may have the heating components included in the cabinet with the cooling components. This type of system is a special application and will be covered later.

Another application, which does not use the furnace blower as the prime air mover, has a fan mounted directly behind the coil to supply the air motion required, Figure 10-3. This type, called a *fan coil unit*, is frequently used in apartment houses, motels or similar places where there is no furnace equipped with a prime

FIG. 10-3. Fan-coil unit mounted in the ceiling of a motel corridor. (Below) Air circulation through the fan-coil blower.

air mover and the space for blower and evaporator is limited. It is ideal where it is not necessary to have a complete duct system or if only one room is to be cooled at a time. Additive heat, in the form of an electric strip heater, can be included with these units to make them a combination heating and cooling unit. In this case the condensing unit will be located at some remote location making it a split or remote system.

LOCATION

As with heating, there are a wide variety of ways in which cooling equipment can be applied to the living space. Models can be combined with up-flo, horizontal or down-flo furnaces. The direction of air is not important as long as the coil is upstream of the blower. In all cases, provision must be made for condensate drains since the evaporator coils will condense moisture out of the air. This condensate must be disposed of.

Condensing units can be located on a slab on the ground, on balconies, on the roof, or any other convenient place where they have access to unrestricted outside air. In selecting a location, keep the length of refrigerant lines short and do not have them rise any great distance above the evaporator. Vertical lines can trap oil and also affect the operation and efficiency of the system.

All brands of air conditioning units will have these same characteristics, even though they may have different shapes and sizes for the various capacities. Fundamentally though, they will all follow the same pattern and operate in essentially the same manner.

INSTALLATION OF A CONDENSER

There will be some variations in the basic installation procedures for a given type of condensing unit, but the following is a basic guide for the installation.

Because of the noise and vibration, condensing units are not set directly on the ground. A small concrete slab, approximately 6 in. thick and slightly larger than the unit, is poured on the ground to support the condensing unit. This slab spreads the weight of the unit and keeps it off the ground.

The slab should be at least 4 in. above ground to assure unrestricted air flow around the base of the condensing unit and to reduce the chance of drawing leaves and dirt into the condenser. Check the grade near the slab. Runoff water, from higher elevations, should not collect around the unit.

Most instructions will recommend maximum distances between the unit and the building wall. In general, the stub connections for the suction and discharge lines should be placed as close to the building wall as is practical to minimize the length of the connecting lines which are exposed, Figure 10-4. However, leave sufficient space for unrestricted air flow to and from the fan and for access to the removable service panels.

Sound-absorbing materials should be placed under the feet or mounting rails of

FIG. 10-5. Sound-absorbing, polyurethane plastic pad is used under the condensing unit. *Courtesy Wencol, Inc.*

the condensing unit to assure the quietest operation possible. An example is a polyurethane plastic pad recently introduced by Wencol, Inc., Figure 10-5.

An alternate, for units with legs, is plastic, dish-like pieces which fit under each leg. These pieces spread the weight on the legs over a larger area to prevent the legs from sinking into the ground.

EVAPORATOR INSTALLATION

Many furnaces will have matching evaporator cabinets that exactly fit the top of the furnace. Where this is true, the evaporator will sit directly over the supply air opening on the flanges provided, Figure 10-6. Usually, two or more sheet metal screws are used to secure the evaporator and its cabinet to the furnace. The space between the flange on the furnace and the evaporator cabinet should be insulated with small fiberglass strips placed completely around the opening. These are

normally provided with the evaporator cabinet.

Where the evaporator does not exactly fit the top of the furnace, most companies supply adapter rails to mount the evaporator. In some cases, a new plenum will have to be built.

The supply air plenum can now be installed directly on top of the evaporator cabinet. This should be insulated and secured with sheet metal screws.

An access panel is needed for future service and maintenance of the evaporator. This rectangular panel is included with the evaporator. Using the panel as a pattern, scribe the outline of the access

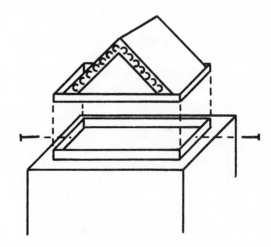

FIG. 10-6. On many furnaces, the evaporator sits directly over the supply air opening.

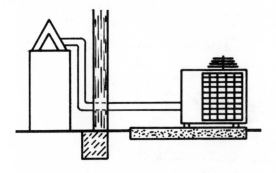

FIG. 10-4. Condensing unit is mounted on a slab and exposed lines are minimized.

panel on the supply plenum, Figure 10-7. Allow about ¼ to ⅜ in. of metal to extend below the supply plenum, so that later it can be screwed into the evaporator cabinet. After scribing, cut a hole approximately ½ to ¾ in. inside of these lines to allow room for attaching the access panel to the supply plenum. Holes should be drilled around the perimeter of the access panel so that it can be tightly secured with sheet metal screws.

If not already available, a ¼ in. hole should be drilled directly above the evaporator, in the leaving airstream, and another one directly below, in the entering airstream. These holes are used to measure the entering and leaving air temperatures so that the temperature drop across the coil can be determined. They are also used as pressure taps to measure static pressure over the coil.

Coils are equipped with a drain connection to remove condensate. This is threaded to accept ¾ in. pipe. The simplest method is to attach a plastic hose to this drain connection and direct it toward a floor drain or outside of the building.

A trap should be installed in the drain line near the unit, Figure 10-8. The trap will prevent air from being drawn back into the unit, through the drain line, and will assure proper drainage of condensate. The trap should be the size of the drain connection or larger. The furnace must be level, or slightly inclined toward the drain, to assure positive runoff.

REFRIGERATION LINES

Suction and discharge lines for an air conditioning system are fabricated in several ways. The normal method, with an expansion valve system, is to fabricate the lines on the job. The refrigerant tubing is joined with either flare connections or by brazing (silver soldering) the lines in the field. Tubing for liquid lines is usually ⅜ in. OD, for up to 3 tons, and ½ in. OD from 3 to 5 tons. The suction line will be from ¾ in. up to 1⅛ in. OD.

The length of run determines the size of the line and the values given would be for an average installation with approxi-

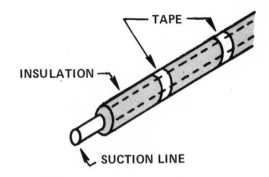

FIG. 10-9. Suction lines are insulated to prevent sweating.

mately 30 ft of tubing. If the suction and discharge lines must be run over 30 ft, then the next larger size piping will be required. The length of the run, as well as the possibility of having to go up, creates a pressure drop through the pipe, which can affect system efficiency. Suction lines should be insulated, particularly if they are exposed to the weather, to prevent sweating, Figure 10-9.

Unsupported suction and discharge lines may vibrate and cause noise within the living space. If it is necessary to run these lines under ceiling or floor joists,

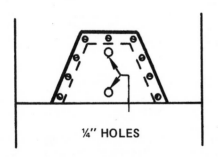

FIG. 10-7. Layout of the evaporator access panel.

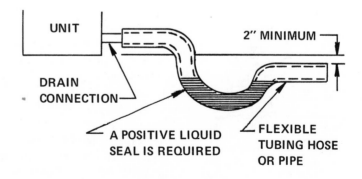

FIG. 10-8. Condensate drain trap prevents air from being drawn back into the evaporator.

easy-to-use hangers are available, Figure 10-10.

If the liquid line is exposed to high ambient temperatures, such as runs on flat roofs or through kitchens or unvented attics, then the liquid line should also be insulated.

Refrigerant piping should never be installed in a cement slab. It is almost impossible to get at the line in the event of a leak, caused by the abrasion of the line as it flexes within the concrete. If the piping is installed under a slab, run it through a sleeve and add insulation at the point where it passes through the slab to the outside.

Horizontal suction lines should be pitched towards the compressor (approximately ⅛ in. for each 10 ft of run) to assure good oil return to the compressor. A certain amount of oil will always

circulate within the system, so deep bends or traps should be avoided in order to allow the oil to return to the compressor.

If the evaporator is 10 ft or more below the compressor, then an oil trap must be provided in the suction line near the evaporator, Figure 10-11. The reason is that the oil returning to the compressor has only the velocity of the return gas to carry it upwards against the force of gravity. As it rises, it tends to adhere to the walls of the tube and run back down, collecting at the lowest point. With a trap, the effective tube area is reduced which increases the velocity of the return gas at this point. The gas will reach sufficient velocity to force additional oil out of the trap and carry it back to the compressor. This eliminates the possibility of oil collecting in the evaporator.

Only refrigeration-grade copper tub-

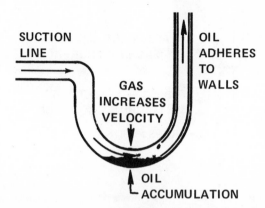

FIG. 10-11. Oil trap in the suction line helps oil return to the compressor.

ing should be used in air conditioning systems. This special tubing is seamless and has been deoxidized, and dried, prior to coiling it into 50 or 100 ft rolls, so that it can be used on the job without additional dehydration. It should be noted that the system must still be dehydrated after installation is complete. This tubing is known as Type L and is available in either soft or hard copper in a number of sizes. It is recommended for use with either flared fittings or brazed connections and is the most popular tubing for refrigeration work. Note that refrigeration tubing is sized by its outside diameter (OD).

Flare fittings or connections are available in a variety of shapes and sizes, including the basic *short knot* female connector, male connectors, union tees and unions. The most commonly used fittings are *short knot* with a female connector. All fittings will have a 45° bevel on the connecting end. The end of the copper piping therefore must be flared at a 45° an-

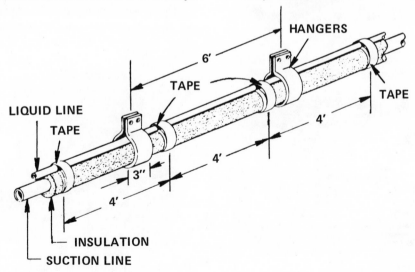

FIG. 10-10. Refrigerant lines are supported by hangers to prevent vibration.

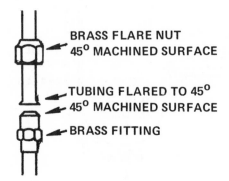

FIG. 10-12. Components of a flare-fitting connection.

gle to mate with this surface, Figure 10-12. The flare nuts draw this flared edge down very tightly to make a completely leak-free joint. It is good practice, when making this joint, to put a few drops of refrigeration oil around the bevelled surface to help create a positive seal.

The tools required for making flare connections are a tube cutter, a flaring tool which makes the 45° bevel at the end of the tubing, and a reamer to remove the burrs and assure a clear, clean surface in the final fitting.

MAKING A FLARE

1. First, accurately measure the amount of tubing required, allowing for the various fittings that go into the piping system.
2. Cut only the amount of tubing required for the immediate connection in order to keep moisture out of the tubing. Use a tube cutter to assure that the cutoff is square and uniform all around, Figure 10-13. Position the tube cutter at the cutoff point and screw it down finger-tight. Too much pressure

will collapse the side of the tube, making it out of round. Too little pressure will allow the cutter to wander around the tubing, making an elliptical or uneven cut. Slowly increasing the tension, as the cutter is rotated, will produce a clean, square cut which will make a good flare.

3. After cutting to the proper length, the tube should be deburred to remove all the turned-in metal resulting from the cut. This can be done either with a reamer, Figure 10-14, or a penknife. Do not use emery cloth or abrasive to remove this type of burr. It is also

FIG. 10-13. Tube cutter assured that cutoff is square. *Courtesy Imperial-Eastman Corp.*

FIG. 10-14. Reamer is used to deburr the tube. *Courtesy Imperial-Eastman Corp.*

important to hold the open end of the tube down so that loose burrs will not lodge in the tubing and later contaminate the system.

4. Set the cut end of the tube in the flaring tool. To assure proper location, use the automatic gauge that is part of the tool. Once the gauge is swung out of position, the flaring tool can be rotated to make the 45° bevel. A small bit of refrigerant oil on the end of the tubing will assist in making a clean bevel. Many flaring tools will burnish this bevel automatically as the cone is withdrawn, so that no further action is necessary. See Figure 10-15.

Be sure the nuts or fittings are placed on the tubing before flaring both ends! For long lengths of tubing, it is sometimes easier to add the connecting nut to the tube prior to flaring it, rather than sliding it from the other end.

FIG. 10-15. Flaring tool is used to add a 45 bevel to the tube. *Courtesy Imperial-Eastman Corp.*

BRAZED CONNECTIONS

Another method of joining refrigeration lines is by silver alloy brazing or silver soldering. Silver alloys have a melting temperature of around 1150° F, which assures a strong, reliable and leak-proof joint. There are many brands of silver solder on the market, each with a different silver content and therefore slightly different melting and flow temperatures. Two commonly used alloys are Sil-fos and Easy-flo, manufactured by Handy & Harman, Figure 10-16.

FIG. 10-16. Easy-flo and Sil-fos silver solders. *Courtesy Handy & Harman.*

Easy-flo alloy contains between 35 and 50% silver and has a flow temperature of about 1100° F. This alloy must be used when brazing copper to steel. (Many compressor valves are steel.)

Sil-fos (sometimes referred to as 95-5) contains either 5% or 15% silver and was developed several years ago for joining nonferrous metals, such as brass, bronze and copper. Brazing temperature or flow point of Sil-fos is approximately 1300° F. Sil-fos will be the less expensive because of its lower silver content.

Fittings designed for a silver solder application will have an inside diameter just slightly greater than the outside diameter of the connecting tubing. This difference in diameters leaves a small capillary space to flow the solder, Figure 10-17. A female fitting will have a shoulder to indicate the proper insertion depth to make a good joint. When the joint is heated to the flow temperature of the solder, capillary action will draw in the solder, completely surrounding the tube and making a sound joint. It is important to

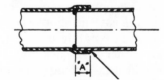

RADIAL CLEARANCE —.002 TO .005 IN.

DEPTH OF INSERT "A" SHOULD BE 3 TIMES TUBE WALL THICKNESS (MINIMUM) GENERALLY MORE FOR ALIGNMENT PURPOSES.

FIG. 10-17. Radial clearance between pipe and fitting aids solder flow.

note that a thinner layer of solder produces a stronger joint. If there is too much clearance between the male and female parts, too much solder will surround the tubing and create a potential source of leakage. Keep in mind that a smaller space between parts also reduces the possibility of corrosion.

Basic soldering techniques can be written down step-by-step, but it is most important for the beginner to practice soldering joints in order to gain firsthand experience of just when a proper solder connection has been made. Three basic things should be constantly in mind in order to braze safely:

1. Use a fan, an exhaust hood and air-supplied respirator, for adequate ventilation.
2. Don't overheat as this increases the amount of toxic fumes.
3. Avoid hot spots. This increases fuming and interferes with good brazing.

The following steps are necessary in order to make a good, leak-proof silver solder joint:

1. Tubing must be accurately measured and proper allowances made for all of the various joints and fittings in the complete tubing system. Depth of insertion into connections should also be allowed for, so that the final tubing system will have a neat and clean appearance. Cut tubing to length, using the roller-type cutting tool described under flare connections. Cut only one line at a time and solder this into the system before proceeding, in order to keep out as much moisture as possible.

When two copper tubes of the same diameter are joined, it is necessary to enlarge the internal diameter of one of these tubes. This is done with a *swaging tool*, Figure 10-18. Some swaging tools are punch-type, where the copper tubing is held in a form and the tube inserted into the open end, up to the top of the angle on the swaging tool, and hit with a hammer. A second is where it is clamped into the swaging bar and a screw-type tool is used, screwing down the swaging tool into the copper to enlarge the inside diameter.

Copper tubing is swaged very easily, but before a swage joint may be made in hard copper, the copper must be softened or annealed. Copper is annealed by heating it to a dark purple color and then allowing it to cool slowly in the air. Care must be taken to avoid the possibility of oxidation. Very hot copper has a high affinity for oxygen and will combine with it to form a scale. Oxidation can be prevented or reduced by heating the tube in the presence of nitrogen. This method will be explained later.

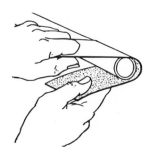

FIG. 10-19 Cleaning tube with emery cloth. *Courtesy Handy & Harman.*

2. Cutting usually leaves a burr on the inside of the tube. This must be removed prior to making the connection, so that small particles of copper will not flow through the system. Remove the burr with either a penknife or a special deburring tool. When deburring, the cut end should be held down so that any particles will fall to the floor rather than back into the tubing. The end of the tubing, both outside and inside, should be cleaned with sandpaper or emery cloth, Figure 10-19. The tube can also be swabbed with a solvent such as R-11, using a small swabbing brush or a cloth. *Do not use carbon tetrachloride as a solvent because it is noncondensible and can cause copper damage.*

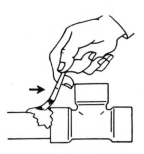

FIG. 10-20. Flux is applied around the circumference of the partially inserted tube. *Courtesy Handy & Harman.*

3. A flux is used to aid the flow of brazing alloy and protect against excessive oxidation. Do not flux the inside the tubing. To properly apply the flux, insert the tube part way into the fitting and then apply the flux around the circumference of the tube, Figure 10-20. Then, insert the tube the rest of the way into the socket and rotate it to make sure that all the surfaces are covered. This will help produce a neat, clean joint. In general, the flux will

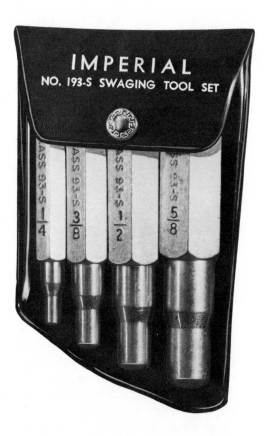

FIG. 10-18. Set of swaging tools. *Courtesy Imperial-Eastman Corp.*

turn to a milky color at about 212°, it
will bubble at about 600° and it will
turn a clear liquid at about 1100° F.

4. The tubing connection, after being cut,
deburred and cleaned, is assembled
and then supported to be sure that it
will not move during the brazing process. Any movement, while the brazing alloy is cooling, can produce a leak,
so it should be supported as firmly as
possible.

5. To reduce oxidation during the brazing
process, the system can be flooded
with nitrogen gas. Nitrogen is relatively inexpensive and can be purchased
almost anywhere. Since the nitrogen
gas replaces the air inside the tubing,
it reduces the amount of oxygen available to combine with the copper. A nitrogen tank, with a pressure regulator
and adapter, is used to slightly pressurize the system. If the end of this system is open, it can be covered with a
balloon, with a small slit in it for an
escape valve, to maintain the slight
pressure required. Brazing in the presence of nitrogen will reduce scale for-
mation and assure much better joints
throughout the system.

6. To form a good joint, both halves of the
joint must be heated above the flow
temperature of the silver solder being
used. An oxyacetylene torch with a
slightly reducing flame—that is, one
with a slight feather on the inner blue
cone, Figure 10-21—is used to bring
the joint and the tube up to the flow
temperature of the silver solder. The
tube should be heated first. Beginning
½ to 1 in. back from the fitting, the

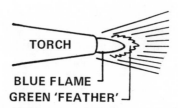

FIG. 10-21. Reducing flame of an oxyacetylene flame has a slight feather on the inner blue flame. *Courtesy Handy & Harman.*

torch must be moved continuously
around the tube to keep from burning
a hole in it. When the tubing comes up
to a dull, cherry red color, and the flux
melts to a milky liquid, then the fitting
should be heated in the same manner,
always keeping the flame moving
around the fitting. When both parts
are up to temperature (the flux should
be a clear fluid on both pipe and fitting), the silver solder is placed firmly
at the point where the tube enters the
fitting. The flame should not be directly in contact with the silver solder.
Figure 10-22 shows the proper application point for the solder on a horizontal
joint and Figure 10-23 shows the point
for a vertical joint. Do not overheat the
pipe below a vertical joint as this will
cause the solder to flow down. Silver
solder will flow due to the heat of the
joint in the tubing and the assistance
of the flux. Make one final pass with
the torch at base of joint. If the fitting
can be rotated while the silver solder is
molten, it will help expel entrapped
gases and flux.

7. Clean and cool the joint with a wet
cloth.

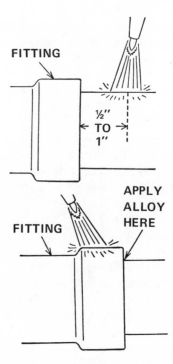

FIG. 10-22. Proper procedure for heating the tube and fitting before (below) applying solder. *Courtesy Handy & Harman.*

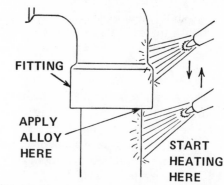

FIG. 10-23. Procedure for heating, then solder-ing a vertical joining. *Courtesy Handy & Harman.*

DISCONNECTING JOINTS

Sometimes it is necessary to disconnect some existing brazed connections in order to repair a damaged part or a leak. Some basic safety precautions should be observed in disconnecting refrigerant lines.

1. Do not apply heat while the line is under refrigerant pressure. The system should be pumped down, so that there is no refrigerant close to the joint to be disconnected, or bled free of refrigerant prior to applying heat to the line. If the line is under pressure when heat is applied, the line can rupture or molten solder may blow out of the joint. Nor is it desirable to have the system under a vacuum since, as the joint is opened, a vacuum will draw air into the system, probably contaminating it with moisture.

2. Flux may be used around the joint as a temperature indicator. When the flux becomes clear, it indicates that the silver solder is molten and that the joint is ready to be disconnected.

3. The system should be slightly pressurized with nitrogen in order to keep air and moisture out when the system is opened.

4. Once the system is opened, all of the openings should be capped immediately to prevent moisture and other contaminants from entering.

5. If there are expansion valves or other devices close to the joint being heated, it is good practice to wrap these with a wet rag so that the heat conducted by the copper line will not reach the valve and damage the diaphragms.

6. All the mating parts must be cleaned thoroughly, exactly as suggested with new piping, so upon reassembly it will form a clean and positive joint.

LINE SETS

Many manufacturers prefer to preassemble their suction and liquid lines so that the problem of cutting, soldering or flaring is eliminated or reduced in the

FIG. 10-24.   Preassembled suction and liquid line sets. *Courtesy Century Engineering Corp.*

field, Figure 10-24. These line sets may be precharged or will contain either nitrogen or refrigerant under positive pressure. In the latter case, the complete system charge will be contained in the condensing unit and the lines must be purged or dehydrated after connection. Line sets come in varying lengths (usually multiples of 5—that is, 15, 20, 25 ft). The length ordered should approximately match the distance between the condensing unit and evaporator. Do not try to shorten line sets with special fittings attached to the lines (some sets have the capillary built into the liquid line and these *cannot* be shortened in the field without destroying its operation). The preferred method is to coil the excess length. Make sure that is is coiled horizontally and sloping toward the condensing unit for easy oil return to the compressor, Figure 10-25.

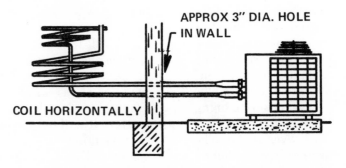

FIG. 10-25.   Excess refrigerant lines are coiled horizontally.

Line sets must be engineered with specially designed couplings or connections at the ends of the tubing, and at the evaporator and the condenser, which will conserve the charge in the condenser during the connection process. There are several of these special valves or connections available that make the connection of line sets possible without losing the system charge. Note that the various fittings are *not* interchangeable and the line set must match the fittings provided by the manufacturer of the coil and condensing unit.

Line sets made by Aeroquip, and others, are precharged with the proper amount of refrigerant required for that length. Lines do not have to be purged or dehydrated after connection. Aeroquip's system uses a fitting or valve which contains a diaphragm that seals the end of the line. As the fittings are connected, a knife blade in the valve cuts through the diaphragm of both the halves. When completely coupled, the diaphragm is cut through and pushed out of the way by the knife edge. Thus, the charge is held in the line set until the connection is secure and no appreciable amount of charge is lost, Figure 10-26.

Installation of these sets is simple:
1. The suction line and liquid line should be routed between the condensing unit and the coil and lined up with the mating fitting.
2. Dust caps and plugs are then removed.
3. Coupling seals and threaded surfaces should be wiped with a clean cloth to prevent any dirt or foreign material getting into the system.
4. The male coupling diaphragm and synthetic rubber seal should be lubricated with refrigeration oil and the two coupling halves threaded together by hand to assure proper mating.
5. Using a pair of properly sized wrenches, the coupling should be tightened down until they bottom or a definite resistance is felt, Figure 10-27.
6. At this point, lines should be marked lengthwise on the coupling union nut and on the bulkhead of the fitting, to indicate the position of the nut. The coupling union nut should then be tightened an additional quarter turn, which can be measured by the offset of the line previously marked on the bulkhead, Figure 10-28.

These couplings can also be tightened down with a torque wrench. The following torque values are recommended.

Foot/Pounds	Coupling Size
10-12	6
35-45	10
35-45	11
50-65	12

FIG. 10-27. Tightening line set coupling. *Courtesy Aeroquip Corp.*

FIG. 10-28. Line set after installation is completed. *Courtesy Aeroquip Corp.*

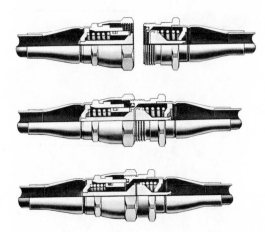

FIG. 10-26.   Aeroquip line set coupling (above) unconnected, (middle) partially connected, (below) fully connected. *Courtesy Aeroquip Corp.*

The couplings may also have a gauge port with a Schrader valve which can be used for gauging and charging the system. Note that the male fitting will be attached to the bulkhead or other part of the unit and the female half will be part of the line set. Aeroquip also makes what they call a *bendable line set*, which is more flexible than straight copper suction lines, and allows a little easier installation. Bendable sets will have the same couplings and valves as other line sets, Figure 10-29.

Another line set system uses a compression fitting to attain an airtight seal. Several manufacturers produce sets of this type. The Aeroquip compression coupling for field assembly includes a gauge port and has a ferrule which tightens down on the tubing, making a clean leak-proof joint, Figure 10-30. Installation of this type of coupling is as follows:

1. Tubing for this assembly should be clean, round, and free of nicks, burrs and scratches with a square cutoff end.
2. The ferrule or sleeve must be clean and free of dents or nicks. The outside diameter is lubricated with refrigeration oil and the long nose is inserted into the coupling body about ⅛ in., Figure 10-31.

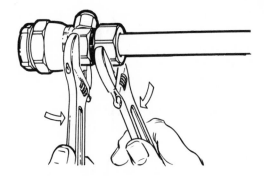

FIG. 10-33. Tightening compression nut.

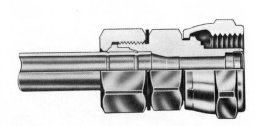

FIG. 10-29. Bendable line sets are equipped with couplings and valves. *Courtesy Aeroquip Corp.*

FIG. 10-31. Inserting ferrule into coupling. *Figures 10-31 to 10-35 Courtesy Aeroquip Corp.*

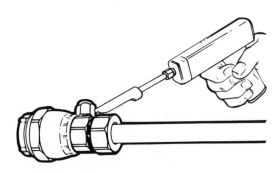

FIG. 10-34. Leak testing the coupling.

FIG. 10-30. Aeroquip compression fitting. *Courtesy Aeroquip Corp.*

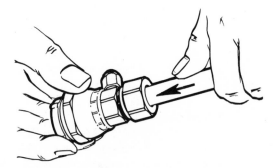

FIG. 10-32. Inseting tube into coupling.

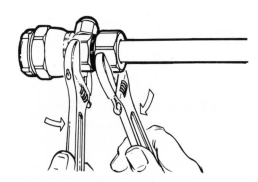

FIG. 10-35. Coupling can be tightened an additional 1/4 turn to stop leaks.

3. The nut is then threaded loosely onto the body.
4. The tubing is pushed through the nut into the body, making sure that the tubing bottoms on the internal shoulder and the bottom of the body socket, Figure 10-32.
5. The coupling body should be held with the wrench and a second wrench used to tighten the compression nut until the end of the nut touches the spacer, Figure 10-33.
6. At this point, the joint can be leaktested by pressurizing with refrigerant and checking with a standard leak detector, Figure 10-34.
7. If leakage is detected, the compression nut can be tightened an additional quarter turn and then rechecked for leakage. If leakage persists, the nut can be tightened up to a maximum of one full turn. The system should be rechecked for leakage after system has been in operation for about ½ hour, Figure 10-35.

Another system which does not require flaring is a valve and compression fitting made by Parker Hannifin, patented under the name Quik-Lok, Figure 10-36. This fitting has a valve which can be *front-seated* (closed), isolating the system from the lines. The valve should be kept front-seated until all connections are secure, the lines purged or dehydrated and, if needed, charge added. It is *backseated* (opened) with an Allen wrench. For a field charged and fabricated system, use the following installation procedures:

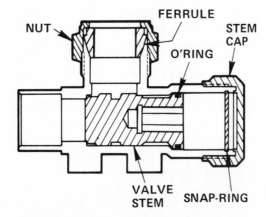

Fig. 10-36. Quik-Lok valve and compression fitting. *Courtesy Parker Hannifin, Refrigeration & Air Conditioning Div.*

1. Remove the rubber plugs or QL plug from the fitting.
2. Remove the nut and lubricate the ferrule with refrigerant oil, which will insure easy and reliable makeup.
3. Install the nut and ferrule on the male fitting. The tapered portion of the ferrule should face the valve or fitting. Loosely thread the nut on the body.
4. Use cleaned copper tubing which has been cut off square and is free of burrs and nicks. Insert the tubing into the nut, making sure that the tubing bottoms on the internal body step.
5. Tighten the nut finger-tight, or to a snug fit using a wrench, and mark this point on both the nut and the body. The nut then should be tightened an additional 1½ turns.
6. The system should be pressurized slightly and tested for leaks. If a leak appears, the nut can be tightened an additional ¼ turn. Once the connec-

tions are made, the normal purging and evacuating procedures should be followed, just as for the regular flare or brazed connection.

This same valve and fitting are also used on precharged line sets. Actually, the line set is merely pressurized with refrigerant and nitrogen, and the charge for about 30 ft of line is included in the condenser. The basic valve is mounted on the condenser and evaporator by the manufacturer and is the same as described under field charge. The tubing comes sealed with airtight rubber plugs which are removed immediately prior to making the connections. If the lines are laid out, approximately in place, the plug can be removed for up to one minute when refrigerant is used as the holding charge, and up to 3 minutes when a mixture of refrigerant and nitrogen is used. This is normally sufficient time to correctly make up this type of fitting.

Installation procedure is exactly the same as recommended for field-charged equipment. Once the connections are made, purge the lines by either using an external refrigerant source, purging through the service ports on the QL valves, or venting a small portion of the system charge to purge the lines and evaporator.

If the system charge is used, the suction valve is opened for about 5 seconds and then closed. At this point the connections should be checked for leaks. If they are tight, the pressure is bled off through the service port on the liquid line valve, which purges the refrigerant and air mixture. Then both valves are opened on the unit and the system is ready to operate.

Most precharged systems will have the complete charge in the high side. The industry practice is to allow, in charging, for a nominal 20 to 30 ft line length. If the line length exceeds 30 ft, then additional charge will probably be necessary for efficient operation of the complete system. The amount required per foot of additional run will be indicated in the manufacturer's installation instructions and this additional charge is added in the normal manner.

Another system is the Primore Mec-Lock Valve, Figure 10-37. To install:

1. Unscrew the coupling nut. Remove the shipping seal plug which holds the "O" ring and crimp collar components. Throw away the shipping seal plug.

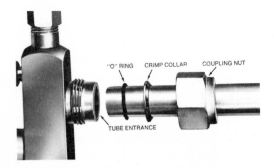

FIG. 10-37.   Primore Mec-Lock valve and compression fitting. *Courtesy Primore Sales Inc.*

2. Be sure the tubing is square, without burrs, gouges or dents and not out of round.
3. Assemble coupling nut on the tubing, open end out to accept the male fitting.
4. Next, assemble the stainless steel crimp collar on tubing with the largest diameter toward open end of tubing.
5. Now assemble the "O" ring on tubing. Be sure to leave enough tubing to fully bottom in the counterbore of the mating, male-threaded fitting.
6. Apply a slight amount of refrigeration oil to the mouth of the male threaded fitting and the "O" ring and crimp collar.
7. Insert tubing into the valve or adapter male threaded fitting. Make sure that the tube is completely inserted into the bottom of the counter bore.
8. Push all components along the tubing and engage threads and tighten coupling nut until it completely *bottoms out*, which insures you that the mechanical coupling joint is complete.

FIELD WIRING

The field wiring of the condensing unit and thermostat for original installation consists of:

1. A disconnect switch at the unit so that it may be turned on and off with a switch close to the unit, Figure 10-38.
2. Line voltage wires from the main circuit breaker or electrical entrance to the disconnect.
3. Line voltage wires from the disconnect to the compressor contactor.
4. Two wires from the thermostat to the blower relay and contactor.

FIG. 10-38.   Safety disconnect. *Courtesy Square D Co.*

5. Line voltage wires from the source to the blower relay and from blower relay to fan control and/or blower motor.
6. Low voltage wires from thermostat to the blower relay and from the blower relay to the compressor contactor.

The size and fusing of the disconnect switch will be called out on the installation instructions and wiring diagram. The disconnect switch is normally not supplied with the unit and must be field-purchased and installed. Fusing for each unit will be specified on the wiring diagram and therefore the size of the disconnect and its proper fusing will be determined by the manufacturer's instructions. The disconnect switch can be located any place convenient to the condensing unit, or attached to the outside of the condensing unit itself. For 240v single phase supply, a 3-lead wire (2 hot and 1 ground) will be run from the main service entrance to the line side of the disconnect switch, Figure 10-39. The line side is marked L_1 and L_3 (sometimes L_2) and the load side to the unit is marked T_1 and T_3 (or T_2). The size of this wire will be called out on the manufacturer's instructions and this should be followed in order to conform with the National Electric Code (NEC). The ground wire is connected to the disconnect box and the two hot leads connected to the fused circuit within the box. *Note that the main circuit breaker should be off, as well as the disconnect switch, and no power should be applied to the unit until all wiring is in place and the manufacturer's instructions advise the application of power.* A 2-lead wire is run from the load side of the disconnect to the compressor contactor, sometimes called the motor controller, and a rubber anti-chafing grommet or electrical connector should be placed on the wire where it goes through the condensing unit cabinet.

When adding cooling to an existing, heating-only system, it will be necessary to replace the thermostat with a heating-cooling thermostat with fan control. This will have a Y terminal for cooling and a G terminal for the fan control. It will also be necessary to provide a blower relay, Figure 10-40, (this may already be provided in the furnace). The blower relay is a single

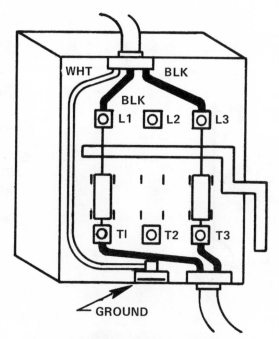

FIG. 10-39. Wiring the safety disconnect.

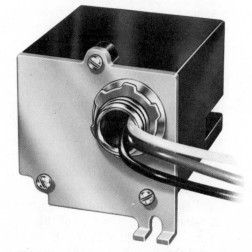

FIG. 10-40. Blower relay may be required when wiring into an existing furnace. *Courtesy Honeywell Inc.*

pole, double throw (SPDT) relay with one set of normally open contacts (NO) and one set of normally closed (NC) contacts plus a coil. This relay allows the blower to cycle with the compressor on cooling and, in the case of multispeed motors, allows the selection of a higher motor speed for cooling than for heating.

Most modern furnaces will have a terminal strip for making the low voltage connections and it will be assumed here that the furnace does have one. (If there is no terminal strip, then wire nut connections can be made in the makeup box.) Wiring for new or add-on units is the same, as the heating connections have already been covered.

The Y and G wires will be run from the thermostat to the Y and G terminals at the terminal block. The Y wire continues to the compressor contactor. A wire from the other side of the contactor is run to C on the terminal block. The G wire goes to

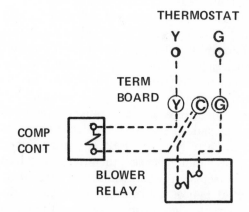

FIG. 10-41. Wiring low voltage circuits into a terminal block.

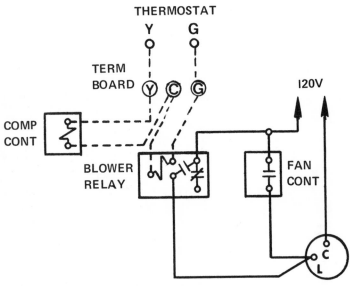

FIG. 10-42. Line voltage wiring with a single speed blower.

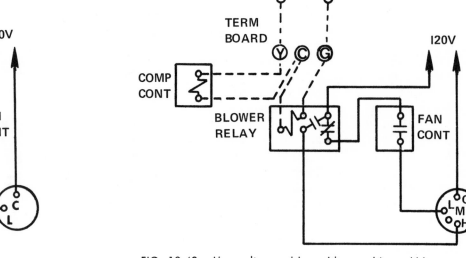

FIG. 10-43. Line voltage wiring with a multispeed blower.

one side of the blower relay. A wire is then run from the other side of the blower relay to C on the terminal block, Figure 10-41.

The common line voltage lead (red on GE-FB32 relay) is connected to the power source. The lead from the NO contact (black) goes to the blower motor. The lead from the NC contact (yellow) need not be used when there is a single speed blower motor unless constant speed blower is required. If so, it can be connected as shown in Figure 10-42. If a multispeed blower motor is used, then the yellow lead connects to the fan control, Figure 10-43. RBM marks the common terminal 4, the normally closed contacts 4-5 and the normally open contacts 4-2. The low voltage coil is numbered 1-3.

In some cases, when adding cooling to an existing system, it is necessary to replace the transformer. The most convenient way of doing this is to add an assembly where the transformer and blower relay are mounted together on a common mounting plate, Figure 10-44. Note that the terminals on the relay are marked to correspond to the terminal markings at the thermostat, so R connects to R, W to W and so forth. The fifth terminal is marked C and the connections to the gas valve and compressor contactor are made here. A fairly typical situation of this type is illustrated in Figure 10-45. Here the primary of the transformer is in the limit control circuit (line voltage) so, if the limit opens, it cuts off power from the trans-

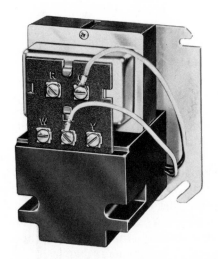

FIG. 10-44. Combination transformer and blower relay assembly. *Courtesy Honeywell Inc.*

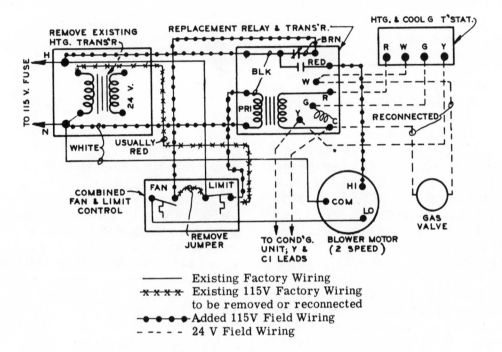

FIG. 10-45. Instructions for substituting transformer/relay for existing transformer.

Legend (in figure):
- Existing Factory Wiring
- ×××× Existing 115V Factory Wiring to be removed or reconnected
- ●●●● Added 115V Field Wiring
- - - - - 24 V Field Wiring

FIG. 10-46. Ratchet wrench used to open and close service valves. *Courtesy Imperial-Eastman Corp.*

former to the entire control circuit. The blower will cycle on the fan control during heating and with the compressor during cooling.

Manufacturer's installation instructions will have step-by-step procedures for field wiring connections and these take precedence over the suggestions given here. The basic connections for an original installation, however, will be essentially as described. Some drawings will show field connections as dotted lines and factory connections as solid lines. Others will show line voltage as solid lines and low voltage as dotted lines, using heavier weight lines to indicate factory and field wiring.

SYSTEM CONNECTIONS

Before going into the methods of evacuating and charging, the valving and connections on the compressor must be thoroughly understood. Valves or fittings are provided on the compressor or line set, making it possible to attach gauge manifolds and thus read the operating characteristics of the system. These valves and fittings also make it possible to evacuate and charge the system without losing refrigerant while attaching or disengaging the hoses. A special tool, called a ratchet wrench, Figure 10-46, is used to open and close the service valves. The wrench will take hold in one direction and slip in the other. This ratchet action allows quick

opening and closing of the valve. The wrench will also stay on the valve stem and not fall off even if the serviceman lets go.

Most compressors will be fitted with a service valve on both the suction and discharge sides of the system. Both service valves will normally be located on the compressor shell. The liquid line valve is frequently located downstream from the drier to allow system pumpdown. The service valve will have an inlet and an outlet for the refrigeration lines and also a smaller port which is called a gauge port. This gauge port is used to check operating pressures and temperatures, and to charge and evacuate the system. If the service valve is completely back-seated, that is, the stem is turned counterclockwise as far as it will go, the gauge port is shut off and there is a clear passage through the valve for the refrigerant. This is the normal operating position when the system is running, Figure 10-47.

After the gauge connection is attached to the gauge port, the valve stem is turned one to two turns clockwise. This is called *cracking* the valve and allows refrigerant to flow past the valve seat and into the gauge port channel. The gauge will now register the operating pressure of the system, Figure 10-48.

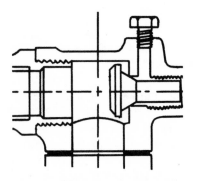

FIG. 10-47.   Turning the stem counterclockwise back-seats the service valve.

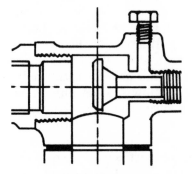

FIG. 10-48.   Turning the stem clockwise cracks the service valve and allows refrigerant to flow to the gauge port.

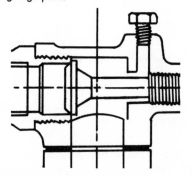

FIG. 10-49.   Turning the step clockwise, as far as it will go, front-seats the service valve.

This valve can also be fully front-seated, that is, turned clockwise as far as it can go, to shut off the flow of refrigerant through the system, Figure 10-49. Condensing units are shipped with the valve in this position so that there will be no chance of leakage of the refrigerant charge during installation.

The gauge port has a Schrader fitting to hold the charge even when the cap is off. This is actually the same type valve as that used to maintain air in a tire. As with a tire, hoses or gauges will unseat the pin when they are screwed down tight.

Some service valves will have two gauge ports. The second port is used in factory processing and does not have a valve stem. If the serviceman is not sure whether the port has a stem, it is good practice to completely back-seat the service valve before removing the cap to determine whether the port is open or has a stem.

A Schrader valve may also be soldered into the line at some convenient place. The Schrader valve functions in exactly the same manner as the service valve gauge port. However, it does not have an adjustable valve stem to shut off the system flow. It has only the valve pin and core which must be unseated when the hose is attached. If the manifold hoses do not have an unseating pin, then a second device must be used, a Superior *unseating coupler.* This attaches to the end of the hose connection and includes the pin needed to depress the valve core on the Schrader fitting. Some units will have both service valves and one or more Schrader fittings. Any of these can be used, depending

FIG. 10-50.   Tool for removing stem from a Schrader fitting. *Courtesy Robinaire Mfg. Corp.*

upon which is the most convenient for the serviceman.

Robinaire has a tool available, Figure 10-50, which will not only unseat the valve, but will also pull the stem out of the way, allowing more flow through the gauge port. This can save considerable time in evacuating and charging.

A common and very necessary tool for installing and servicing air conditioning equipment is a charging manifold, Figure 10-51. This manifold has been developed to

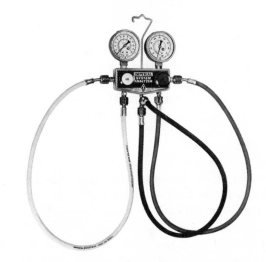

FIG. 10-51   Charging manifold. *Courtesy Imperial-Eastman Corp.*

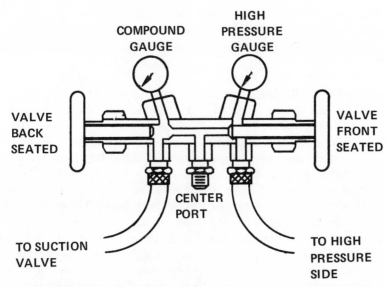

COMPOUND GAUGE

HIGH PRESSURE GAUGE

VALVE BACK SEATED

VALVE FRONT SEATED

CENTER PORT

TO SUCTION VALVE

TO HIGH PRESSURE SIDE

FIG. 10-52. Controlling flow through the charging manifold. Suction side is open to the compound gauge and center port. High side is closed.

do a number of different jobs without changing hose connections into the system. The charging manifold has a compound or low pressure gauge which has a scale from 200 pounds down to 30" of mercury vacuum. It therefore can read a vacuum as well as a positive operating pressure. A second gauge is normally located to the right of the compound gauge as you look at the manifold. This is a high pressure gauge which should go up to at least 400 psig.

When the gauge manifold valves are front-seated, (turned full clockwise), the center port is completely closed off from the gauges and gauge ports, Figure 10-52. The gauge port, however, is open to the system and the operating pressures will

be indicated. Opening (turning counterclockwise) either valve opens that side of the system to the center port.

Three hoses are attached to this manifold: one on each side and directly below the gauges, plus a third one in the center. In addition, there are two hand valves, one at each side, which allow one or more of these hoses to be opened into the system. Some will have an angle fitting on the center hose for connecting a fourth hose.

The hose to the compound gauge is attached to the suction service valve gauge port. The hose to the high pressure gauge is attached to the discharge or liquid service valve gauge port or liquid line Schrader fitting. The center hose is attached to either the refrigerant supply cylinder or

the vacuum pump. Where the fitting has two hose connections to the center port, one hose goes to the refrigerant cylinder and one to the vacuum pump.

There are charging stations available which include additional piping and hand valves. With these, the vacuum pump, refrigerant cylinder or manometer can be selected merely by turning a hand valve, without changing the hose from one to the other. These charging stations are often mounted on a small wheeled cart for ease of movement from truck to the condensing unit, Figure 10-53.

The hand valves on the gauge manifold will open each side of the system to the center hose and, if both are open, both sides are open to the center base. The closed position for these valves is the clockwise position. They should be closed firmly to assure that no refrigerant will leak out of the system.

FIG. 10-53. Typical charging stations. *Courtesy Airserco Mfg. Co.*

CONNECTING THE SYSTEM

If a charging stand or a dual hose connection for the center port is not available, then it will be necessary to attach the hoses alternately to the vacuum pump and refrigerant cylinder. Since the first step in a field-installed, uncharged, new installation is to evacuate the system, the center hose from the gauge manifold should be connected to the vacuum pump. The fitting is screwed down handtight. The left hand hose (as you face the manifold) is attached to the gauge port on the suction valve of the compressor and screwed down tight. The right hand hose is connected to the gauge port of the liquid line valve or Schrader valve in the liquid line and screwed tight, Figure 10-54. The left hand, or compound gauge, can now show vacuum or pressure on the suction side of the system, and the right hand gauge will show pressure on the discharge side.

This procedure for attaching hoses is satisfactory on *initial* installation, prior to the time the system has been evacuated, but will *not be satisfactory* after the system has been dried, due to the possibility of getting air or moisture into the system. The proper procedure for attaching the hoses to the system is as follows: Prior to attaching the refrigerant drum, unscrew and take off the hoses to the suction and discharge valves. The center hose should be attached to the refrigerant drum and both manifold valves should be closed. Crack the refrigerant drum valve to establish pressure in the hose. The hose to the discharge side of the system should be very loosely attached to the discharge gauge connection. Then, the hand valve on the manifold to that side of the system should be cracked. This will allow some refrigerant to flow through the hose and purge any air in the line. After the line is purged, the hose should be held tightly against the gauge port and screwed down securely and quickly. This procedure should be repeated on the suction side so that both hoses will be completely purged of air and moisture before being opened into the refrigeration system. When attaching the hoses, the service valve should be completely back-seated so that the gauge port does not open into the system.

Once a system is charged and under pressure, the hoses must be removed quickly and positively or the charge can be lost. To remove the hose, hold the metal angle tightly against the gauge port as the nut is loosened. When the nut is free, remove quickly so no charge will be lost. The valve should be back-seated when removing the hose to prevent loss of charge.

EVACUATING THE SYSTEM

All new installations, where tubing is fabricated on the job, need to be evacuated to a deep vacuum in order to be sure that all noncondensible gases and moisture are removed from the system before charging. It is also necessary to evacuate any system that has been opened to the atmosphere, since moisture or other noncondensibles can get into the system while it is being serviced.

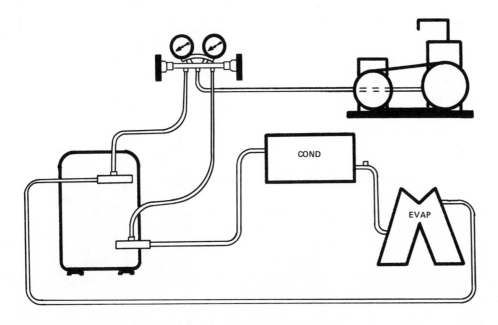

FIG. 10-54. Connecting the charging manifold prior to evacuation.

Manufacturers go to considerable trouble and expense to reduce the moisture content in their components to the absolute minimum. As explained under refrigerant, moisture in a system can make it inoperative in a short time. It would be an expensive shame, therefore, to override their efforts by contaminating the system with moisture during installation. Proper evacuation assures a dry, uncontaminated system.

Vacuum is measured in inches of mercury. The absolute possible vacuum is 29.999 in., referred to as 30" barometer or .001" mercury absolute. Measurements this small become somewhat unwieldy, so vacuum is often expressed in the metric system. A *torr* (one millimeter) or a *micron*, which is 1/1000 of a torr, are the metric units of measurement. There are about 25 torrs in an inch, so a reading of .001" Hg (mercury) absolute would be .001 x 25 or .025 torr Hg. Converted to microns, this becomes 25 microns.

The vacuum pump is required, in order to get a deep vacuum, to pull at least 29¼" of mercury. In order to remove the water in a system, it must boil into a vapor. At ordinary ambient room temperature (approximately 70°) water will not boil until the vacuum exceeds 29¼". Many vacuum pumps will only take the system down to 27 or 28". At 28", water will boil at 100° so, unless heat is added, no moisture will be removed. Since they will not remove moisture, these pumps are not recommended for refrigeration servicing. A vacuum pump need not have a pumping capacity of over 1 to 3 cfm, since the size of the gauge port hole will not permit more

air to pass, Figure 10-55.

During evacuation, the vacuum should be measured by using a very accurate U-tube manometer, Figure 10-56, or an electronic gauge to assure good dehydration. This is important because the vacuum gauge on the manifold is not divided into small enough increments to read accurately. A Bourdan type vacuum gauge should *never* be used. Most vacuum pumps will only have one inlet connection and no provision for attaching the manometer. A small T connection should be inserted at the inlet to the pump with one leg going to the manometer and the other legs going to the system.

With the vacuum pump connected, and *both* manifold service valves open, the pump can be started and the system evacuated. It is a good practice to stop the pump at least once during the evacuation period to see if the vacuum will hold at that point. If it will not, there is a leak in the system which has to be located and repaired before an absolute vacuum can be obtained. When the system has been evacuated down to 29¼" or better, wait 5 to 10 minutes to see if this vacuum will hold. If it does not, there are leaks in the system or the system is not completely dry.

Turn off the pump and front-seat the manifold valves. Then, remove the hose from the pump and attach it to the refrigerant drum. After purging the hose, the vacuum should be broken by opening the refrigerant drum and manifold valves. Allow the system to pressurize to 2 or 3 lbs. Check for leaks and repair if present. Then close both manifold valves, discon-

FIG. 10-55. High efficiency vacuum pump. *Courtesy Robinaire Mfg. Corp.*

FIG. 10-56. U-tube manometer. *Courtesy Dwyer Instruments, Inc.*

nect the refrigerant drum from the center port and open the suction line valve slightly, purging the system down to 0 psig. The system should then be evacuated a second time, through the suction service valve, until the vacuum within the system does not rise above 29.7" of mercury.

CHARGING THE SYSTEM

Air conditioning systems can be charged with refrigerant in either the vapor phase or in the liquid phase, depending upon the amount of charge required, the valving arrangement on the unit, and the outside temperature. For new installations or where a large amount of charge is required, the system would normally be charged with liquid since this is faster. If only a small quantity is required, then it likely would be charged with vapor through the suction side, since the system is then operating more closely to normal conditions and can balance out the new charge more quickly.

Some units, because of their valving arrangement, can be charged only in liquid or only in vapor form and this will determine how the system is charged.

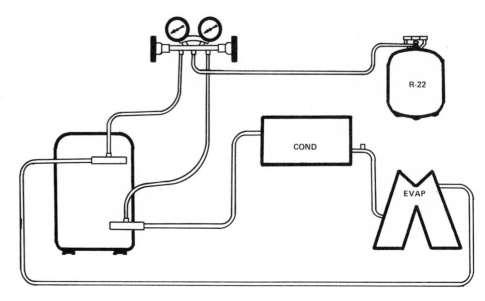

FIG. 10-58.   Charging with refrigerant in the vapor phase.

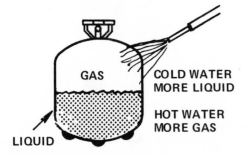

FIG. 10-57.   Adjusting the liquid/vapor ratio prior to charging.

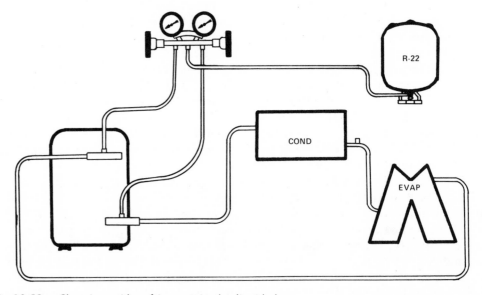

FIG. 10-59.   Charging with refrigerant in the liquid phase.

Finally, keep in mind that a refrigerant cylinder will contain both gas and liquid, with the gas rising to the top of the cylinder and the liquid staying at the bottom. As the outside ambient temperature increases, there will be considerably more gas in the cylinder than liquid. If the temperature drops, there will be considerably more liquid in the cylinder. Thus, it may be necessary to either heat or cool the cylinder with water in order to get the amount of gas or liquid required for charging, Figure 10-57. *Never use an open flame* or similar means to warm the refrigerant as this can result in too rapid expansion of gas within the cylinder.

If the system is to be charged with a gas, the cylinder is placed upright. If the system is to be charged with a liquid, it will be turned upside down. All cylinders have a guard rail around the valve. This is designed to act as a stand for the cylinder when it is in the upside-down position.

In preparing to charge the system, the hose to the compound gauge should be purged and attached to the suction service valve, and the hose to the high pressure gauge should be purged and attached to the discharge service valve, Figure 10-58. If the unit is to be charged with vapor, then the refrigerant drum is connected to the center hose and refrigerant charged through the low pressure or suction side of the system. Where it is necessary to have a full charge, charging can be speeded up by front-seating the suction service valve until there is approximately a full charge. At that time, the service valve is opened and the compressor started and run for about 5 minutes, to allow the system to stabilize.

When liquid charging a system, the manifold may be connected in the same way as for vapor, but the refrigerant drum is turned upside down and the high pressure side of the manifold is opened to the refrigerant drum, Figure 10-59. Again, where a complete charge must be added, it will speed up the charging procedure if the liquid line valve is front-seated until the charging is almost complete.

Systems can also be charged by weight, but this requires a very accurate scale, since the charge in many systems cannot vary more than 1 or 2 ounces. The attachments to the suction and discharge service valves are made in the same way as with vapor charging and the suction service valve is front-seated. Initially, the charge is allowed to enter the system from the pressure of the refrigerant in the drum. As this slows down, the compressor can be started to speed the introduction of the charge.

Another method of charging by weight, which claims to be accurate to + or -¼ oz, is a cylinder charging system offered by Robinaire and others, Figure 10-60. Directions for use come with the cylinder. Basically, the procedure is to fill the cylinder with refrigerant, following the instructions. The amount is shown by a dial shroud that is calibrated in pounds and ounces. The proper charge is fed into the system through the gauge manifold in the same manner as charging from a refrigerant drum *except* that the charge is controlled from the cylinder. System charge is checked in the normal way.

Since it is most important that the amount of charge in the system be accurate, particularly in units without receiver

FIG. 10-60. Charging cylinder premeasures the charge for greater accuracy. *Courtesy Robinaire Mfg. Co.*

tanks, a method of checking the charge must be used to tell the serviceman when the correct amount has been added.

CHECKING REFRIGERANT CHARGE

The most accurate and widely used method of checking refrigerant charge is by a chart which relates suction pressure to outside temperature and then to discharge pressure. These charts are made up for *each specific unit* by the manufacturer, who determines the values, under controlled laboratory conditions, using a system with the correct amount of charge, Figure 10-61. The outside temperature is normally indicated by a curve drawn through the center of the chart. The chart is entered on the left with the discharge pressure read on the gauge. The serviceman moves horizontally across the chart, along the discharge pressure line to the ambient or outside conditions curve, and then straight down to the suction pressure expected. When the system is charged correctly, the suction pressure and discharge pressure will be extremely close to the points noted on the chart. Some manufacturers use ambient temperature as the vertical scale, with discharge pressure as the horizontal scale, and draw the curve for suction pressure. However, the basic use of the chart is essentially the same.

A second method of checking the refrigerant charge is to block the condenser until the head or discharge pressure reaches about 300 psig. Suction pressure is held to about 50 psig by adjusting the suction valve. When these two pressures can be brought into equilibrium, the system contains the proper amount of charge.

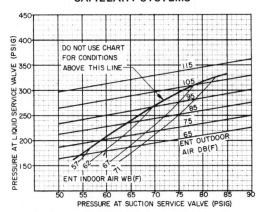

CAPILLARY SYSTEMS

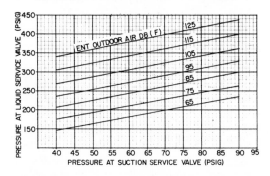

THERMAL EXPANSION VALVE SYSTEMS

FIG. 10-61. Charts used to determine proper refrigerant charge.

This is not always a satisfactory field method because it is difficult to hold the suction pressure exactly at 50 psig and the ambient temperature can affect the readings.

Some manufacturers equip the condenser with either one or two tubes which can be opened and checked. These tubes are located at the point where the refrigerant changes from a gas to a liquid. Thus, when the top tube sprays gas and the bottom tube sprays liquid, the system is close to being properly charged. Where a single tube is used, the system is properly charged when the tube sprays liquid rather than gas. This system is less accurate than the chart method, but it is used by several manufacturers.

Another method of checking charge is by a sight glass located in the liquid line between the condenser and the drier, Figure 10-62. A sight glass is just what its name implies: that is, a small glass bubble that allows the observation of the refrigerant as it is passing through the liquid line.

FIG. 10-62. Sight glass indicates when there is insufficient charge. *Courtesy Imperial-Eastman Corp.*

The refrigerant at this point should be almost 100% liquid. Therefore, if bubbles are observed in the line, the system is undercharged. After the system has been operating for about 5 minutes, the refrigerant in the sight glass should be totally free, or almost so, of any bubbles when it is properly charged. Some sight glasses will confirm that a system is low on charge by colored lines which appear in the sight glass. Many sight glasses also indicate, by a color change, whether the system is wet or dry.

UNDERCHARGED SYSTEM

If any of the above checks indicates that the system is undercharged, or if the gauges are showing low suction pressure and/or low head pressure, then refrigerant must be added. Refrigerant is added using exactly the same procedures as employed during the original charging. It is recommended that when charge is added to the system, it be done in the vapor phase.

OVERCHARGE

Without some accurate guidelines for variations in operating pressures and temperatures due to outside ambient conditions, it is easy to overcharge a system. If the ambient temperature is below the normal rating conditions of about 95°, then pressures will also be reduced. Reading subnormal pressures on the gauges will lead the serviceman into overcharging the system.

Overcharge of a system can be indicated by high head (discharge pressure) or high suction pressure, or it may be indi-

cated by the suction line sweating.

If the system becomes overcharged, turn off the valve on the refrigerant tank, remove the hose from the refrigerant tank, and then slightly crack the low side manifold valve. This will allow the refrigerant to bleed out of the center hose. Bleed a little bit of refrigerant at a time. Then close the manifold valve to allow the system to stabilize. After a few minutes, check to see if the pressures, temperatures or other indications of overcharge are corrected. Continue to bleed charge until the indications show a correct charge.

TESTING FOR LEAKS

All new installations should be checked for leaks prior to complete charging. As indicated earlier, leaks can be detected during the evacuation process, since a leak in the system allows air to enter, and raise the pressure, when the system is under a vacuum. If a system has lost all or most of its charge, then the system must be pressurized again, up to approximately 150 lbs minimum. This can be done by adding refrigerant, using the usual charging procedures. Or it may be pressurized with nitrogen, which is less expensive than refrigerant. Nitrogen will also leak faster than R-22 and is not absorbed by refrigeration oil. Nitrogen cannot, however, be detected by a leak detector or halide torch.

Due to the explosive pressures of nitrogen, it should never be used without a pressure regulator on the tank.

On the other hand, leaks in a system pressurized with refrigerant can be spotted with a halide torch or an electronic

leak detector which will pick up an extremely small refrigerant leak. Therefore, this discussion assumes that the system is pressurized with either all refrigerant or a mixture of nitrogen and refrigerant.

If a system has been operating for some time, the first check for a leak might be made visually. Refrigerant will carry with it a small quantity of oil. Therefore, traces of oil at any joint or connection would lead the serviceman to suspect that the refrigerant is leaking at that point.

One rather simple and inexpensive method is to use soap bubbles to check for a leak. Any solution of water and soap can be used. Some servicemen use the soap solution sold in novelty stores for blowing soap bubbles. The solution should be liquid and not be shaken prior to application. This reduces, as much as possible, the number of bubbles in the solution when it is applied. The soap solution is applied to all joints and connections in the system. A small pinhole leak will be identified by bubbles forming in the soap solution around the leak.

Another method employs the halide torch which heats a copper element, Figure 10-63. This flame will change color in the presence of a halogen refrigerant (R-12, R-22, etc.). The flame is turned down as low as possible and will usually be a steady blue color. The hose, which supplies primary air to the flame, is passed around each joint and fitting in the system. If refrigerant is present, it will be introduced into the flame, along with the primary air, and turn the flame green or greenish purple. Caution is advised in using a halide torch. In the presence of an

FIG. 10-63. Halide torch with sensing tube. *Courtesy BernzOmatic Corp.*

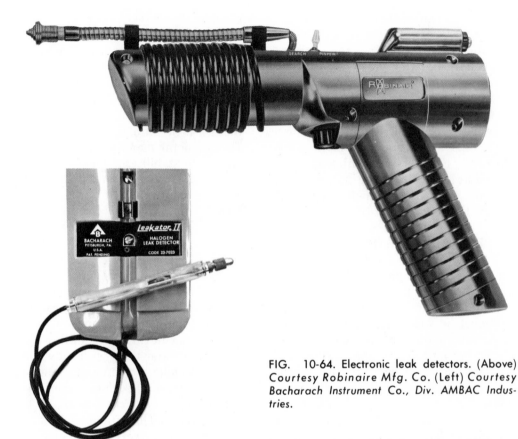

FIG. 10-64. Electronic leak detectors. (Above) *Courtesy Robinaire Mfg. Co.* (Left) *Courtesy Bacharach Instrument Co., Div. AMBAC Industries.*

open flame, halogens can form phosgene which is a deadly toxic gas. Therefore, the halide torch should only be used in well-ventilated areas.

Because of this hazard, the halide torch is probably the least desirable method of leak detection. Further, it is not as efficient as an electronic leak detector. It is, however, inexpensive, accurate and widely used.

In recent years the electronic leak detector has been developed and perfected.

This is unquestionably the most efficient and easiest method of checking for leaks, Figure 10-64. There are various types of electronic leak detectors available. Generally speaking, they are all portable, most are lightweight, and consist of a box with several switches and a probe or *snifter*. The detector is turned on and the probe passed around all of the fittings and connections in the system. The leak will be indicated either by movement of a pointer on the dial on the face of the machine, by a

buzzing sound which is easily audible, or an indicator light that goes out if a leak is discovered.

In all cases, whenever a leak is found in the system, the system must be *pumped down* and the leak repaired before proceeding with final charging and operation.

Some systems have valve arrangements so that only a portion of the system needs to be pressurized. If it is suspected that the source of the leak is on either the high side or the low side, then only this side needs to be *pumped down*. Generally speaking, it is best to pressurize the entire system so that it is thoroughly checked for leaks before proceeding.

After it has been determined that the system has the correct charge, and does not leak, the compressor service valve is back-seated and the gauge manifold hand valves closed. Hoses are removed from the compressor by holding the angle fitting down tightly, unloosening the threads and removing quickly so as not to lose charge. The gauge caps are replaced and the system is ready for normal operation.

SYSTEM PUMPDOWN

Many systems have valving arrangements which will allow repair of leaks in the low side piping and evaporator, removal of the expansion valve or drier and complete replacement of the compressor without losing the refrigerant charge. This is accomplished by *pumping* the entire charge into the condenser and holding it there (new split system units are often shipped this way). This is called *pumping down* the system.

The system must be equipped with a suction service valve, a discharge service valve and a liquid valve in order to pump it down without losing the charge, Figure 10-65. Units equipped with line set fittings such as Parker Q-Lok locate the liquid line valve between the drier and the evaporator so the drier cannot be changed without losing the charge. Line sets with quick connect couplings such as Aeroquip *must have* a multiple shot (Series 5500) coupling to save the charge.

The system must also have an adjustable low pressure control which can be

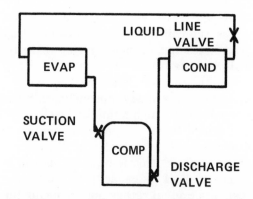

FIG. 10-65.   A system equipped with suction and discharge valves and a liquid line valve may be pumped down.

turned down to zero prior to pumpdown. Nonadjustable low pressure controls should be set at about 5 lbs so the compressor can be pumped down to this point before it will cut out. This allows the saving of some of the charge but the balance will have to be bled out. The low pressure control can be jumpered to *start* the compressor, but jumpering is not recommended to keep the compressor running.

Since it only takes a few minutes to pump the compressor down, it may be convenient to stop the compressor quickly, when it pumps down, to avoid damage. This can be done by wiring a toggle switch at any convenient point in the low voltage control circuit. The compressor operation can then be controlled at this point. To pump down the system:

1. Set the toggle switch to *off* or have unit disconnect off.
2. Set the room thermostat well above room temperature.

3. Put the fan control in the *on* or *cont* position.
4. Set the low pressure control at zero.
5. Attach a gauge manifold set to the unit.
6. Crack (turn clockwise) the suction and discharge valves 1 to 2 turns to open the gauge ports.
7. Completely front-seat (turn clockwise) the liquid line valve (or disconnect the multishot quick connect).
8. Start the compressor by the toggle switch or unit disconnect and allow it to run until the suction pressure comes close to zero. Stop the compressor.
9. After the first pumpdown cycle, the pressure will usually increase. Start the compressor again (jumper the low pressure switch, if necessary, to start) and repeat the pumpdown cycle until the suction pressure holds at 1 or 2 psig. Keeping a slight positive pressure in the system prevents air and moisture from being drawn in if a leak exists in the evaporator.
10. Completely front-seat the discharge service valve.
11. Repairs or change-out in the compressor or low side can now be made.
12. If system has been opened, evacuate the system and install a new drier.
13. Open suction and discharge valves (turn counterclockwise). Slowly open the liquid line valve (or reconnect the quick connects).
14. Reset the low temperature control and remove toggle switch (if used).
15. Check operating temperatures and pressures. Add charge if needed.

ADJUSTMENTS TO SYSTEM

In order for any air conditioning system to operate properly, it must have the correct amount of air passing over the evaporator coil. The methods of changing the amount of air delivered by the blower have already been discussed. These included adjusting the pulley on a belt-drive blower and, for a direct-drive blower, changing either the speed controller or tap-wound motor connections. The amount of air passing over the coil can be determined by a measurement of its static pressure drop. Whether this is sufficient air to provide a good cooling system is determined by measuring the temperature drop across the coil. All air conditioning systems are assumed to deliver between 400 and 450 cfm of air per ton of rated refrigeration capacity. In other words, a 2-ton system should have between 800 and 900 cfm across the coil, and a 3-ton system between 1200 and 1350 cfm

of air across the coil. The amount of air across the coil and its temperature drop are related but do not correlate exactly. This is because the temperature drop across the coil can be affected by the humidity on a given day. For instance, if the basement has a high humidity level, it will be difficult to get the proper temperature drop across the coil, even with the correct amount of air. This is because the coil is using much of its capacity to reduce the moisture content rather than reducing the sensible temperature of the air. Under these circumstances, the coil is actually functioning correctly even if the temperature drop is less than expected. Removal of humidity is just as important to the comfort conditions in the house as reducing the sensible temperature of the delivered air. Thus the temperature drop across the coil can vary quite a few degrees. Under normal conditions, the temperature drop across the coil will be between 18 and 20° F.

Most units will have ¼ in. holes drilled in the plenum, one directly above the coil and one directly below the coil, Figure 10-66. These are for measuring the entering and leaving air temperatures or for measuring the static pressure across the coil. These holes may be covered up by the insulation on the inside. If so, an awl or sharp-pointed device can be inserted into the hole to clear it for insertion of the instruments. If these holes are not present, they can be drilled. However, the serviceman should be extremely careful not to drill into the evaporator. The hole should

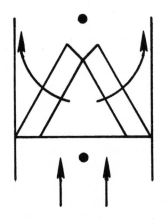

FIG. 10-66. Points used to measure static pressure.

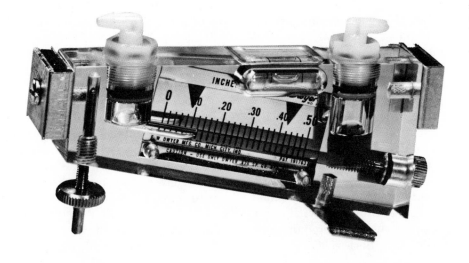

FIG. 10-67.   An inclined water gauge manometer is used to measure static pressure. *Courtesy Dwyer Instruments Inc.*

be located on the center line of the evaporator.

The pressure drop across the coil, or the static pressure, is measured in inches of water gauge with an inclined manometer or draft gauge. *Static pressure measured across a dry coil, that is, without the refrigeration circuit operative, is most accurate. However, many manufacturers give the range of static pressure over a wet coil and some give both.* If the refrigeration circuit has been in operation, it is better to wait 10 to 15 minutes with the blower on, before taking these measurements for a dry coil.

An inclined manometer, Figure 10-67, allows direct reading of static pressure across a coil To take a static pressure reading with an inclined manometer, proceed as follows:

1. Remove the manometer from its box or carrying case.
2. Open each connector for the tubing by turning one turn counterclockwise.
3. Level the manometer with either the adjustable foot or by placing it against a steel surface and holding it with the magnet, generally supplied with the manometer. A spirit level is built into the manometer to show when the gauge is absolutely level.
4. Adjust the scale so that the zero reading is exactly at the point where the fluid in the manometer shows. Attach one hose to the connector at the lower part of the manometer and place the other end in the upper hole in the evaporator. The second hose is attached to the connector at the higher end of the manometer and the opposite

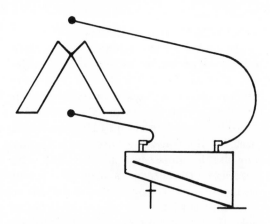

FIG. 10-68. Hose connections to the water gauge manometer.

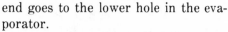

end goes to the lower hole in the evaporator.

At the point where the hoses enter the cabinet, place a small dab of Permagum on the outside of the cabinet so that the hole will be sealed completely and an accurate reading obtained, Figure 10-68.

5. Turn the blower on (reduce thermostat setting well below room temperature to keep air conditioning system off, if the measurement is to be taken on a dry coil) and measure the static pressure as shown by the meniscus in the manometer.

Many manometers have a mirror behind the fluid to allow accurate reading in the scale. By aligning the mirror image and the actual fluid, parallax is eliminated and an accurate

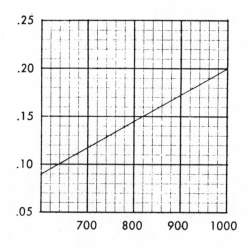

FIG. 10-69. Typical static pressure chart furnished by manufacturer.

reading can be obtained.

The reading in inches of water gauge can then be compared to the manufacturer's instructions, which will show the amount of cfm passing across a given coil, Figure 10-69. The instructions will show readings for specific coils and the proper coil must be selected. Note whether the values given are for a wet or dry coil. The readings are often given in a spread (.20 - .22) since it is usually difficult to read an exact number with these small divisions.

The curve or table for static pressure assumes a furnace plenum, as in Figure 10-70A. If the plenum is opened, as in Figure 10-70B, it can result in a drop of .05" with the same quantity of air.

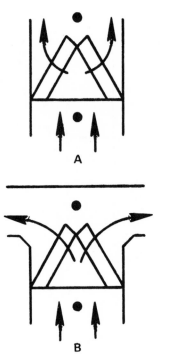

FIG. 10-70. Static pressure tables assume a closed plenum (A). Open plenum (B) causes a pressure drop.

If the reading obtained on the manometer is below the readings given in the table for the desired cfm, then the blower speed should be increased and the pressure tested again. If the reading is above the range selected, then the blower speed is greater than that desired and should be reduced.

The amperage draw of the motor should be checked, when it is operating at a speed that produces the required static pressure, and this reading compared with the rated amperage on the motor nameplate. If the reading is close to the amperage on the nameplate, then it can be as-

sumed that the motor is not overloaded at the speed selected for proper air over the evaporator.

TEMPERATURE DIFFERENCE

In order to check the temperature difference or temperature drop across the coil, turn on the unit and allow it to operate for 5 or 10 minutes or until the system stabilizes. If two thermometers are available, place one in each of the two holes used for the static pressure check. If only one thermometer is available, take the reading on the entering side of the coil and then move the thermometer to the other hole and take a reading on the leaving side. If more convenient, the entering air can be measured in the return air duct, close to the furnace.

A properly operating coil, with the required amount of air across it, will have a temperature difference between these two readings of about 18° F. As mentioned earlier, the temperature drop can vary with the humidity in the immediate area. Therefore, the system is not necessarily malfunctioning if the temperature drop is less than the 18° .

In general, if the temperature difference is lower than 18° , then the blower speed is too fast. If it is higher than 18° , the blower speed is too slow. However, this should be considered in view of the results of the static pressure test. If the blower speed is about right, the lower temperature drop is likely due to high humidity. There may, however, be something else wrong in the refrigeration system and checks for this will be discussed later.

STARTING KITS

Most units of 3 tons and under are equipped with capillary tube metering devices. The capillary tube allows the system to unload (stabilize pressures) on the off cycle. As a result, the compressor motor will not be starting against a wide pressure difference and minimum starting torque is required. Under normal conditions, this presents no problem. However, if the supply voltage is low, then the compressor will not be able to start, due to the reduced starting torque. When this occurs, the compressor will go off on the internal overload. If a low voltage condition persists, a *start kit* can be added to the system to increase the starting torque and avoid nuisance trips. A *start kit*

FIG. 10-71. Start kit is composed of a start capacitor and a potential relay.

consists of a start capacitor and a potential relay, Figure 10-71. The start capacitor helps the motor start and the relay removes the capacitor from the circuit when the motor is up to speed.

Wiring diagrams are furnished with the start kit and the basic wiring is as shown in Figure 10-72. Note that T_1 goes to the start capacitor and also the run capacitor (marked side) then continues to the *run* terminal of the compressor. From the start capacitor it goes to No. 1 on the relay. The No. 2 terminal of the relay has a connection from the run capacitor which then goes to the *start* terminal on the compressor. The No. 5 relay terminal and the common terminal of the compressor go to T_2.

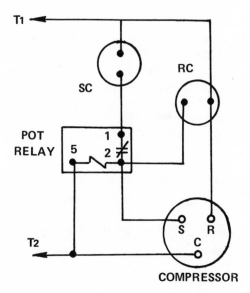

FIG. 10-72. Start kit wired into the circuit.

MAINTENANCE

Basic maintenance for blower motors, including adjustment and lubrication, was covered under furnaces. The same procedures apply to a cooling system check.

In addition, the evaporator should be checked for any clogging by dirt or lint. If there is any accumulation, the coil can be cleaned with a wire brush, taking care not to damage the fins of the coil.

The condensate drain should be checked for free flow. This can be done by pouring a cup of water in the drain pan to see if it runs off quickly.

All tubing joints and connections should be checked for signs of oil which would indicate a refrigerant leak.

Furnace filters should be checked and changed if clogged or dirty.

The condenser should be examined to make sure that all surfaces are free of leaves or any other obstructions. If not, it should be cleaned. The condenser can be cleaned with a garden hose or wire brush if leaves are particularly difficult to remove. The condenser fan motor, if not permanently lubricated, should be oiled approximately once a year with a few drops of automotive SAE No. 10 nondetergent oil, or that specified by the manufacturer. The compressor compartment should be visually checked for evidence of oil leaks as should the connections to the suction and liquid line. Electrical connections should be checked for tightness and proper contact.

Note that in making any of these checks and operations, the condenser should be turned off at the disconnect switch to avoid unforeseen start-up.

Questions

1. What components make up the low side of a refrigeration system?
2. What components make up the high side of a refrigeration system?
3. What is a self-contained system?
4. Where can condensing units be located?
5. How are condensing units mounted on the ground?
6. Why is a condensate trap important?
7. What are the four basic ways to connect refrigeration lines?
8. How should horizontal suction lines be pitched and why?
9. When is an oil trap necessary in the system?
10. What kind of tubing is used for air conditioning?
11. Name the tools required for making a flare connection.
12. What is the melting temperature of silver brazing alloys?
13. How are the parts to be silver soldered assembled together?
14. What draws the solder around the tube to make a good joint?
15. What are the three basic safety considerations when brazing copper tubing?
16. Give the seven steps for good brazed connections.
17. What is done with the excess tubing in a prefabricated line set?
18. How does a compression-type fitting differ from a special fitting such as the Aero-quip Coupler?
19. Where does a precharged system store the refrigerant charge?
20. What electrical connections are required for the installation of a condensing unit?
21. When adding cooling to an existing heating system, what additional parts will have to be supplied?
22. What are the basic service valves on a compressor and where are they located?
23. Back-seating a service valve does what?
24. What is meant by cracking the valve?

25. When the valve is fully front-seated, what is the situation as far as the system is concerned?
26. What is a Schrader valve?
27. What is a charging manifold?
28. When the gauge manifold valves are turned full clockwise, is the center port open or closed? Are the gauge ports open or closed?
29. Where is the refrigerant drum attached?
30. Where is the vacuum pump attached?
31. What must be done to the hoses prior to attaching them to the system? Why is this done?
32. Why is the system evacuated?
33. Why is it necessary to draw a deep vacuum?
34. How is the vacuum measured during evacuation?
35. What can be checked during the evacuation process? How?
36. What would be the normal way of charging a new system? Why?
37. If only a small amount of charge is required for an existing system, how would this be charged? Why?
38. How can you change the proportion of gas to liquid in the refrigerant cylinder?
39. Name three ways of checking for the proper refrigerant charge in a system.
40. Name four ways to check the system for leaks.
41. What is meant by system pump down?
42. How much air per ton is required for proper operation?
43. What is a normal temperature drop across the coil?
44. How is static pressure measured?
45. What information can you get from the static pressure reading?
46. When is a start kit required?
47. What does a start kit do?
48. List the major maintenance items for a refrigeration system.

Chapter 11
HUMIDIFIERS

Thus far, this book has been concerned with how air is controlled and how it is moved from place to place after passing through the heating or air conditioning system.

It has been established that air has weight, volume and some moisture content. While these qualities are not easily seen, and their quantities are constantly changing, they can be measured. In the case of moisture content, a major change, either up or down, can be sensed by the human body. Further, most materials, wood, leather, plaster, paper, rugs, etc., will absorb or lose water depending upon the moisture content of the surrounding air. They will shrink as they dry out and swell as they absorb water.

A low moisture content will cause nasal passages and the throat to dry out. Many doctors suggest that this is one cause of the common cold. Under similar conditions, furniture and woodwork will crack or warp and rugs will wear out much more quickly because the dried fibers will break. Another common symptom of low moisture is static electricity, which gives one a shock when an object is touched.

On the other hand, a high moisture content usually makes one feel dull and not up to par. Furniture will swell, doors will stick or not close, and water will condense on windows and run down, ruining wood sills, drapes or rugs.

Thus, it is most important, in a total comfort system, to control the moisture in the air within comfortable limits during both the heating and cooling seasons.

MEASUREMENT

The amount of moisture in the air is measured in *grains*. A grain is a very small unit since it takes 7000 grains to equal one pound of water.

Moisture in the air is called humidity or, more properly, *relative humidity* (rh). The ability of the air to hold moisture will vary with its temperature, and relative humidity is based on a comparison between the amount of moisture actually in the air and the maximum amount of moisture that air can hold at that temperature (100% saturated). For example: air at 70°, if 100% saturated, can hold 110 grains of moisture per pound of air. If the 70° air actually has 55 grains of moisture per pound, then it is said to be at 50% rh (55/110 = .50 or 50%).

As the temperature of the air increases, its ability to hold moisture also increases. If the same 70° air is heated to 85°, it can now hold 185 grains of moisture. If no moisture is added to the air, then its rh *drops* to 30% (55/185 = .297 or 30%). Thus, increasing the temperature of the air, without adding more moisture, *reduces* the relative humidity.

Going the other way, if the temperature of the air is reduced, then its ability to hold moisture *decreases* and the relative humidity will *increase*. Cooling the 70° air down to 55°, without adding moisture, will increase the relative humidity to 85% because 55° air, when saturated, can only hold 65 grains of moisture (55/65 = .846 or 85%).

This is the reason that moisture content is always stated as relative humidity, since it is totally dependent upon the temperature of the air.

The temperature at which air reaches its saturation point (100% rh), or is holding all the moisture it can, is called its *wet bulb* temperature. The actual temperature of the air, or sensible temperature, is called the *dry bulb* temperature. Dry bulb temperature will be higher than the wet bulb until 100% rh is reached and then they will be equal.

Relative humidity can be measured directly by a device called a sling psychrometer, Figure 11-1. The sling psychrometer is a simple instrument that simultaneously measures the wet and dry bulb temperatures. It consists of two thermometers, mounted side by side, on an arm that is free to rotate. A small sock or wick is soaked with water and placed around the bulb of one of the thermometers. The psychrometer is held by the handle and the two thermometers whirled rapidly in the air for about a minute. The moisture from the sock will be evaporated into the air, due to the movement of air across it, and this will tend to cool the wet bulb. The dryer the air, the more water will be evaporated and the lower the wet bulb reading will be.

FIG. 11-1. Sling Psychrometer. Courtesy Bacharach Instrument Co., Division of Ambac Industries.

By placing the wet bulb temperature and the dry bulb temperature readings opposite one another on the slide included with the psychrometer, relative humidity can be read directly as a percent.

INFILTRATION

During the heating season, warm, moist indoor air is constantly escaping, through doors, windows, even the roof and walls, and being replaced by cold, dry, outside air. This process is called infiltration and the amount of cold air entering the structure is measured in air changes per hour. An air change is, by definition, the quantity of entering outside air that displaces all of the inside air in the total cubic contents of the structure.

The number of air changes that occur in an hour is dependent upon the construction of the house. Construction is defined by the National Warm Air Heating & Air Conditioning Association as follows:

Tight   A structure with insulated walls and ceiling, with a vapor barrier, tight-fitting storm doors and weather stripping. Air infiltration is about ½ an air change per hour.

Average   A structure with insulation in the walls and ceiling, loose-fitting storm windows and doors with possibly some vapor barrier. Air infiltration is about one air change per hour.

Loose   A structure with no insulation, no weather stripping, no storm doors or windows and no vapor barrier. Air infiltration will be about 2 air changes per hour.

Therefore, depending upon the house construction, large volumes of cold, dry outside air are constantly coming into the house and circulating through the heating system. This cold outside air cannot hold much moisture. Even at 90% rh, its moisture content, in grains, is very low. For instance: if outdoor air at 0° and 90% rh is heated to 72° indoor temperature, without adding moisture, its new rh will be only 5%, see Table 11-1.

People living in a house will add moisture to the air, as will cooking, showers, dishwashing, clothes washing, etc., but this is usually not enough to offset the very low rh of the heated outside air. These moisture-producing activities are usually concentrated at certain times of the day, so they do not provide a continual source of moisture.

The only satisfactory answer is a humidifier whose capacity is matched to the infiltration in a particular house. This humidifier becomes part of the heating system and introduces additional moisture into the duct system. Some will introduce a fixed amount of water per day or per hour, while others are controlled by a humidistat that monitors the moisture content in the house and turns the humidifier *on* or *off* to maintain a preset rh. The humidistat can be located on the wall in the living room, next to the thermostat, or it can be located in the return air duct. Use of a humidistat is preferred because it will hold the rh at a preset level. This is important because it is not desirable to raise the rh too high during the winter.

TABLE 11-1. Relative Humidity Conversion Chart

OUTDOOR RELATIVE HUMIDITY	-10°	-5°	0°	+5°	+10°	+15°	+20°	+25°
100%	3%	4%	6%	7%	9%	11%	14%	17%
95%	3%	4%	5%	7%	8%	10%	13%	16%
90%	2%	4%	5%	6%	8%	10%	12%	15%
85%	2%	4%	5%	6%	8%	9%	12%	15%
80%	2%	4%	5%	6%	7%	9%	11%	14%
75%	2%	3%	4%	5%	7%	8%	10%	13%
70%	2%	3%	4%	5%	6%	8%	10%	12%
65%	2%	3%	4%	5%	6%	7%	8%	11%
60%	2%	3%	3%	4%	5%	7%	8%	10%
55%	1%	2%	3%	4%	5%	6%	8%	9%
50%	1%	2%	3%	4%	4%	6%	7%	9%
45%	1%	2%	3%	3%	4%	5%	6%	8%
40%	1%	2%	2%	3%	4%	4%	6%	7%
35%	1%	2%	2%	3%	3%	4%	5%	6%
30%	1%	1%	2%	2%	3%	3%	4%	5%
25%	1%	1%	1%	2%	2%	3%	3%	4%
20%	1%	1%	1%	1%	2%	2%	3%	3%
15%	+%	1%	1%	1%	1%	2%	2%	3%
10%	†%	+%	1%	1%	1%	1%	1%	2%
5%	+%	+%	+%	+%	+%	1%	1%	1%
0%	0%	0%	0%	0%	0%	0%	0%	0%

OUTDOOR TEMPERATURE

Courtesy Research Products Corp.

COMFORT LEVELS

Most people feel quite comfortable when the rh is between 30 and 60%, but there are some other complicating factors which sometimes make this difficult to achieve during the heating season.

CONDENSATION

Moist air has another characteristic which is called its *dew point*. The dew point is the temperature at which the moisture in the air will condense out and turn to water. For example, air at 70° and 50% relative humidity will have a dew point temperature of 51°. If this air comes in contact with an exposed window, that has a surface temperature somewhere below 50°, the water in the air will condense out and run down the window. Note that the surface temperature of the window, due to conduction from the outside air, can be considerably below average room temperature. If the temperature of the window is 32°, or at the freezing level, then the moisture will freeze, forming ice and frost on the window. This is not desirable since the sun's heat will melt this frost and cause it to run down onto the sills and window structure. This dripping condensate can easily cause structural damage around the window frames and may also soil drapes and floor coverings.

Frost or condensation on large window surfaces is a sure indication that the humidity level is set too high. If condensate is evident, the humidistat should be set back until the windows are clear. During extreme winter months, the humidity level is always a compromise between the amount of humidity that the homeowner desires and the amount of humidity that

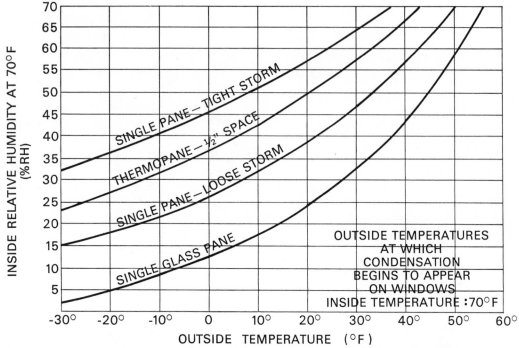

FIG. 11-2.   Temperatures at which condensation appears on window glass. *Courtesy Research Products Corp..*

will cause frost or condensate to form. Under severe weather conditions, it is probably a wiser choice to reduce the humidity until the outdoor temperature rises. These options should always be pointed out to the homeowner.

The temperature at which condensation will form depends upon the type of glass used in the windows. Moist air will condense on single pane glass at much higher outside temperatures than on double glass or thermopane. The attached chart, Figure 11-2, shows these condensing points. The chart shows that, at an outside temperature of 0°, condensate will form on single pane glass when the

relative humidity inside is about 13%. If thermopane is used, condensate will not form until the relative humidity goes above 37%. Thus, with insulated glass, a higher rh can be maintained within the house without condensation. For this and other reasons, most manufacturers recommend holding the indoor rh within certain limits, depending upon the outside air temperature. Their suggestions are shown in Table 11-2.

SIZING AND APPLICATION

Humidifiers are usually rated in terms of gallons of water per hour, or per day, that they will deliver into the house

TABLE 11-2.

OUTSIDE AIR TEMPERATURES	INSIDE RELATIVE HUMIDITY OF INDOOR AIR AT 70 DEGREES F
-20° or below	not over 15%
-20° to -10°	not over 20%
-10° to 0°	not over 25%
0° to +10°	not over 30%
+10° or more	not over 35%

duct system. The total gallons needed will depend upon the cubic capacity of the house, its construction, the winter outside temperature, the desired inside rh and heating system temperature.

Precise calculation is difficult, because compromises are required due to possible condensation and steps in the basic sizes of humidifiers offered by manufacturers. Further, the outside and inside temperature and rh conditions are constantly changing, so there is never a stable set of conditions. However, most manufacturers will provide a set of guidelines to determine the proper size and type of humidifiers for a given house, using the variables discussed above. One suggests that a one gallon/hour humidifier will handle a loosely-constructed 11,000 cubic foot house, an average 22,000 cubic foot house and a tightly-constructed 44,000 cubic foot house.

TYPES OF HUMIDIFIERS

There are basically three types of humidifiers, each with many things in common. All will introduce water vapor into the house air stream at predetermined rates. All require a source of water and most require a drain connection.

FIG. 11-3.   Pan-type humidifier with vertical plates. *Courtesy Skuttle Mfg. Co.*

Some require a humidistat, with additional field wiring, and a blower or rotor motor. These are the basic types:

PAN TYPE HUMIDIFIERS

The basic pan-type (or vaporizing) humidifier uses heat from the air, or from an external source, to vaporize the water. The vaporized water particles then mix with the heated air supply. A simple pan of water placed on a radiator was once considered adequate humidification. But, to do even a passable job in a small house, a 10 ft x 10 ft pan would be required. This is hardly practical today.

One way to increase the capacity of a pan-type humidifier is to place a rack in the pan and insert porous, mineral wool plates into the rack, Figure 11-3. The bottoms of the plates rest in the water and absorb water by *wick* action until the entire plate is wet. This design increases both the surface area and the humidifying capacity. The water level in the pan is maintained by a float valve. The pan, with

the plates, is inserted into the furnace plenum where the warm supply air picks up moisture from the plates as it passes over them.

The pan-type humidifier is installed by cutting a hole in the plenum and mounting the whole assembly on the furnace. Water supply and drain are then connected. There is no electrical requirement since the float valve maintains a preset, constant water level.

A variation of the pan-type humidifier has an electric heating element immersed in the water. The heater boils the water and creates vapor in the form of steam. Immersion heaters are often used with portable humidifiers and with baby vaporizers.

ATOMIZING HUMIDIFIERS

An atomizing humidifier converts water to very small droplets before introducing them into the air stream. This can be done in several ways, Figure 11-4. One method is to use a circular wheel or cone that rotates at a fairly high speed. Water

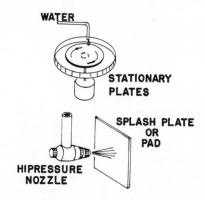

FIG. 11-4.   Two types of atomizing humidifier. *Courtesy Research Products Corp.*

is fed onto the rotating wheel where centrifugal force converts it into small droplets. This method is often used with self-contained humidifiers.

Another method uses a high-pressure spray nozzle to blow water onto a pad or splash plate. This type is attached to the furnace plenum with a bypass to the return air duct. Warm air will pass across it, due to the air pressure differential between the supply and return sides of the system, and pick up and carry moisture to the return side of the duct system.

WETTED ELEMENT HUMIDIFIERS

Wetted-element humidifiers are possibly the commonest type used with forced, warm air systems. They employ a porous medium that is wetted by either passing water through, from top to bottom, or immersing it constantly in water. Air is passed over the media by either an air pressure differential or a self-contained fan. Keep in mind that, with all systems working on air pressure differential, the *direction* of air flow will be *from* the warm air plenum *to* the return air plenum, regardless of where the unit is mounted. This air current picks up moisture from the media and carries it into the house duct system.

One example is the rotating drum type humidifier. A cutaway of this model is shown in Figure 11-5. The unit consists of a drum covered with a polyurethane evaporator pad that is rotated very slowly (1 rpm), through a pan of water, by a 24v motor. Water in the pan is kept at a constant level (about 1-3/8 in.) by a float valve included with the assembly. The bottom part of the pad is immersed in the water

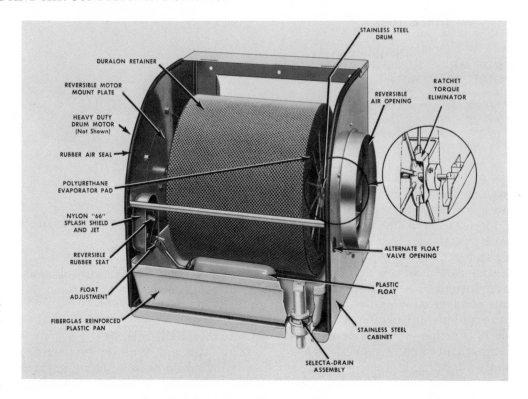

FIG. 11-5.   Cutaway view of a rotating drum type humidifier. *Courtesy Skuttle Mfg. Co.*

and the entire pad is soaked as the drum rotates.

A needle valve and saddle brackets, Figure 11-6, are supplied to tap into the house water system. A copper tube, with compression fittings, connects the needle valve to the humidifier.

A drain is provided to remove excess water from the pan. This drain also allows a periodic flushing action which is most important in removing accumulated mineral deposits.

The humidifier is inserted into the air supply through a hole cut in the furnace

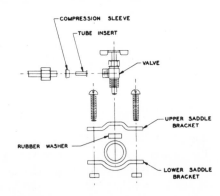

FIG. 11-6.   Needle valve and saddle brackets. *Courtesy Skuttle Mfg. Co.*

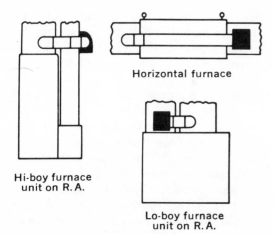

Horizontal furnace

Hi-boy furnace
unit on R. A.

Lo-boy furnace
unit on R. A.

FIG. 11-7. Typical humidifier installations. *Courtesy Skuttle Mfg. Co.*

plenum. The manufacturer recommends that this particular model be placed in the return air plenum, for better efficiency, but it can also be mounted on the warm air plenum. A bypass tube is connected to the side of the humidifier and run to the warm air plenum. Figure 11-7 shows some typical installations. Air is drawn across the pad due to the air pressure difference across the two sides of the system. The air picks up moisture and returns to the house duct system.

The operation of the drum is controlled by a humidistat wired in series with the 24v drum motor. The humidistat

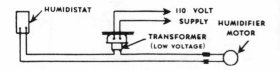

FIG. 11-8. Humidistat is wired in series with the humidifier motor.

can either be mounted on the wall, near the thermostat, or in the return air duct. If mounted in the duct, care should be exercised to be sure it is out of the path of any radiant heat. Figure 11-8 shows the wiring for this unit. With this arrangement, the humidifier will operate independently of the furnace cycle and provide additional moisture whenever the humidistat senses a drop in humidity. Best results can be obtained by keeping the furnace blower on at all times to move the greatest amount of air over the humidifier. When wired correctly, the humidifier operates regardless of whether the burners are on or off.

NOTE Top-of-the-line furnaces from many manufacturers will have a terminal board in the furnace to make either 24v or 115v humidifier connections.

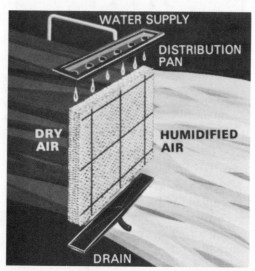

FIG. 11-9. Functional view of top-fed, wetted element humidifier. *Courtesy Research Products Corp.*

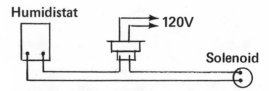

FIG. 11-10. Humidistat in series with water-feed solenoid.

Another way to use the wetted element principle is to mount the evaporator pad vertically and feed water from the top, Figure 11-9. Water must be evenly distributed across the entire top surface and the medium must allow the water to pass slowly from top to bottom to maintain a uniformly wet surface. Air is then forced through the wetted medium.

One often-used material is asbestos which is formed into a corrugated shape, with the open ends facing the direction of air flow. Most of these pads indicate the proper direction of air flow by an arrow on the pad. This type of unit also requires a drain connection, since the water that is not evaporated will drip down to the bottom of the unit. Since there is no reservoir of water, the water flow must be controlled by a humidistat and allowed to flow only when the humidistat senses a drop in humidity. This is done by wiring a solenoid water valve in series with the humidistat. When the humidistat contacts close, they will energize the solenoid coil, opening the valve and allowing the water to flow into the top of the unit. Figure 11-10 shows the basic wiring for a 24v system. With this wiring, the humidifier will operate independent of the furnace cycle.

One such unit is shown in Figure 11-11. This unit mounts directly on the warm

FIG. 11-11. Vertical, wetted-element humidifier. *Courtesy Research Products Corp.*

FIG. 11-13. Vertical, wetted-element humidifier equipped with a fan. *Courtesy Research Products Corp.*

air is circulated by a separate fan built into the humidifier. With this system, no bypass pipe is needed and the unit mounts directly on the furnace plenum, Figure 11-13. Water supply and drain lines are required, as in the previous version.

Air bypass openings flank the pad on both sides, allowing the fan to draw warm air directly from the plenum and then push it, through the evaporation media, back into the plenum, Figure 11-14.

Fan motors and solenoid water valves are 115v and are controlled by either a line or low voltage humidistat. In most applications, the humidifier operates only when the furnace is running. The basic wiring, Figure 11-15, shows the humidistat in the line voltage circuit. If the humidistat is in the low voltage circuit, then a relay must be added, Figure 11-16.

There are many variations in the application of these basic types of humidifiers, depending upon a given manufactur-

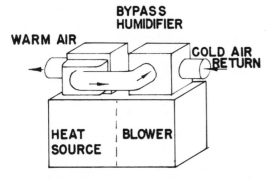

FIG. 11-12. Vertical, wetted-element humidifier is installed with a bypass pipe. *Courtesy Research Products Corp.*

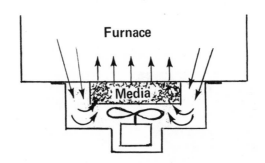

FIG. 11-14. Functional view of air circulation in fan-equipped, vertical wetted-element humidifier.

11-12. Warm air is drawn through the wetted pad by the air pressure differential between the two sides of the system. The moving air picks up moisture, by evaporation, and returns to the duct system.

Still another variation of this principle has the pad in the vertical position and feeds water to it through a solenoid operated water valve, as before. However, the

air side of the plenum and has a bypass pipe running from the face of the unit to the return air side of the system, Figure

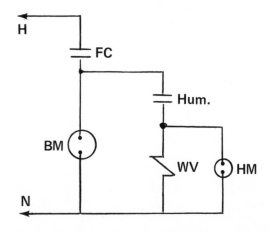

FIG. 11-15. Humidistat wired into line voltage circuit.

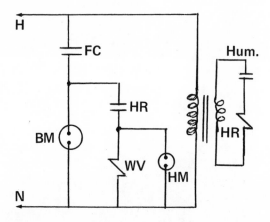

FIG. 11-16. Humidistat wired into low voltage circuit.

Class of Hardness	Grains Hardness per Gallon	Area of U.S. by percent
Low	3 - 10	30
Average	10 - 25	55
High	25 - 50	15

er's design. Some humidifiers (usually used in areas with water having a low mineral content) recirculate the water within the unit and do not require a drain. Various media are used to present the moisture to the air stream and different piping connections may be suggested for best results. All designs, however, follow these principles.

MAINTENANCE

All water, except distilled water or rainwater before it reaches the ground, contains certain mineral deposits held in solution. When the water is evaporated, these minerals in the water remain as dust particles or as a residue that can and does cause problems.

Water is classified by its *hardness:* that is, the number of grains of mineral matter it contains per gallon. This will vary considerably, according to the area, and can even vary greatly within the same city. One breakdown is as follows:

The harder the water in a given area, the greater the problem with mineral deposits. Because of the water, no type of humidifier will function properly in some areas. Water hardness test kits are available and are quite simple to use.

Note that water softeners *add* minerals to the water and, while most humidifiers can use either hard or soft water, the use of softened water does not change the basic problem.

The obvious problem is that the minerals remaining after the water is evaporated will circulate through the system and settle as dust in the house or will remain in the humidifier and clog pads, drains and other operating parts. The first and most important maintenance procedure is to thoroughly clean all operating parts and change media pads regularly. Extreme water hardness may require changing pads and cleaning the system every 4 to 6 weeks. Polyurethane media can be removed, washed in soap and water, and replaced. A 50% vinegar and water solution is a good solvent for most mineral deposits, and parts should be cleaned with it. Many companies offer special cleaning solutions for their units which help to solve this problem.

Most motors used in humidifiers are protected by a plastic enclosure or other shield to reduce the accumulation of min-

eral deposits on operating parts. The motors need oiling every season. Electrical connections, such as solenoid valves, are vulnerable even though they are protected with plastic sleeves. Float valves can acquire a mineral coating and either function erratically or not at all. Even models with regular flushing cycles, to minimize mineral buildup, can end up with clogged drains.

Spray nozzles are particularly vulnerable to mineral buildup. The best solution to this is to alternate nozzles every year. The out-of-service nozzle should then be soaked in vinegar until the next season.

Another common humidifier malfunction is caused by a mineral buildup at the needle valve in the main supply. This valve should be checked each year as it can cut off all water to the unit.

Electrical checks should also be made on the transformer, humidistat, fan motor, solenoid or drum motor to determine if each is operating properly. In the main, however, maintenance and service on humidifiers will concern itself with the consequences of mineral deposits on the operating parts.

Questions

1. Name some signs of low humidity.
2. Name some signs of high humidity.
3. What is the unit of measurement for moisture content of air?
4. Define relative humidity.
5. What happens to rh when air temperature is increased?
6. What is saturation point?
7. What is used to measure rh?
8. How does it work?
9. What is infiltration?
10. How is house construction defined?
11. What is the normal rh level?
12. What is dew point?
13. Why is condensation a problem?
14. How are humidifiers rated?
15. Name three basic types of humidifiers.
16. On what principle does each operate?
17. What is the direction of air flow on a pressure differential type humidifier?
18. How is the main water connection made?
19. Do all humidifiers require a drain? Why?
20. What medium is used for evaporating pads?
21. What does a humidistat do?
22. Is the humidistat wired into the low or line-voltage circuit?
23. Does a drum type humidifier cycle with the furnace?
24. Does a power type humidifier cycle with the furnace?
25. What is the importance of water hardness?
26. Does a water softener improve humidifier performance? Why?
27. Describe the most important maintenance procedure for a humidifier.

Chapter 12
ELECTRONIC AIR CLEANERS

Over the past few years, the subject of air pollution has received considerable attention and caused considerable concern for the future. At the national level, a new agency of government, the Environmental Protection Agency, was formed to deal with these problems.

What is meant by air pollution? Pollution means that the air contains additional substances over and above its normal makeup of 21% oxygen, 78% nitrogen, .03% carbon dioxide and minute traces of other elements. When large concentrations of other substances accumulate in the air, it is no longer in its pure form and is considered to be polluted.

These pollutants are added to the air in many ways. Natural occurances, like earthquakes, windstorms or forest fires, add dirt, dust or ash to the air. Pollutants can be residues from industrial plants, they can be the products of combustion from utilities, incinerators, coke plants, automobiles and trucks, or they can evaporate into the air from polluted water. As the need for more energy has expanded over the past few years, the amount and types of contaminants in the air has increased rapidly. Whatever the source, contaminants likely will be present for some time.

This chapter will seek to identify the substances that are of personal concern and show how they can be eliminated or controlled within the immediate home environment.

MEASUREMENT OF POLLUTANTS

Pollutants take many forms and may occur as gases, oxides, hydrocarbons, particles and vapors. Some are merely irritating, making one's eyes water or clogging nasal passages. Some, like pollen, viruses, and bacteria, are carried into the lungs or circulatory system and cause colds, allergies, or more serious problems. A few pollutants can be fatal to plants or small animals.

This chapter will be limited to particle pollution. While particles make up only about 13% of total air pollution, they do act as *carriers* of other dangerous contaminants, like sulphur dioxide, and they can be controlled.

A particle is defined as (1) an object having definite physical boundaries in all directions and (2) those objects with diameters ranging from 0.001 microns to 100 microns. Particles can be either solid or liquid.

A particle is extremely small and its diameter is measured in *microns*. A micron is 0.000001 of a metre. It would take 25,400 microns to equal one inch, or about 200 microns to reach across the dot made by a sharp pencil. A particle of fine sand would measure about 100 microns.

The visible particles (down to about 10 microns) make up only about 10% of the total number of airborne particles, the rest are invisible. Particles under 5 microns will stay suspended in the air for a considerable length of time and be circulated constantly through the furnace and its filter system. Particles under 1 micron, for all practical purposes, will stay suspended in the air indefinitely. The oily or greasy particles (.01 to 1.0 microns) will eventually come in contact with either a vertical or horizontal surface and adhere to it. When this happens, they stain and soil walls, drapes, and so forth, which then require cleaning or replacement.

FILTER SYSTEMS

All furnaces are equipped with some form of air filter. These are usually a mechanical impingement type filter that has a fiberglass, polyurethane, or metal media. The media is oiled or coated on one side to collect and hold airborne particles as they pass through the filter. These filters are effective for particles down to 5 to 10 microns. Most house air systems are designed for 3 to 4 air changes per hour, which means it takes between 15 and 20 minutes for each air change. In this time period, some of the 10 micron or larger particles will settle out as dirt or dust in the house and others will be trapped by the filter. The larger particles, that settle out as dust, are visible and are easily wiped off with a dustcloth or vacuumed up. The smaller particles will pass freely through the filter and be recirculated through the house.

As it collects dust and lint, the mechanical filter does get somewhat more efficient and picks up some of the smaller particles. However, this greater efficiency increases its resistance to air flow. Therefore, it has to be changed or cleaned periodically, in order to maintain proper temperature and air circulation. For all practical purposes, the mechanical filter will eliminate particles down to about 10 microns. This includes dust, lint, pollen, plant spores and some other substances like fly ash, Figure 12-1.

The electronic air cleaner is designed to eliminate the smaller particles measuring from 10 microns down to .01 microns in size. These include viruses, tobacco smoke, oil smokes and similar particles. A good EAC will eliminate 95% of the particles in this size range.

Many air cleaners will have an additional filter made from activated charcoal. This is to remove some of the gaseous contaminants that cause odors. An EAC, by itself, will not normally remove odors, so the combination makes a very satisfactory system. Care should be taken that the charcoal filter does not come in contact with the high voltage wires, because some charcoal filters are combustible and can smoke or catch fire.

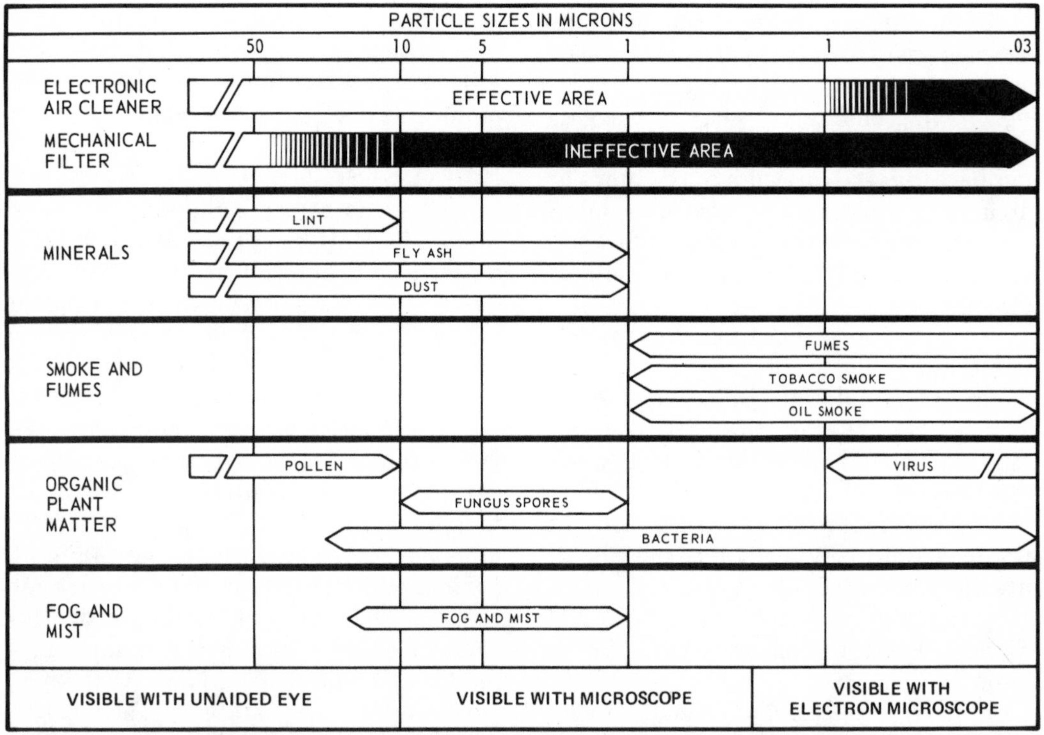

AIR CLEANERS

An air cleaner, as opposed to an electronic air cleaner, does not have a power pack or any electrical connections. An example is Research Products' Space-Gard cleaner. Basically, this system uses a media folded in accordian fashion, approximately 8" deep. This media permanently entraps contaminants and the manufacturer claims it will not unload. (Unloading means that when the filter is full of trapped particles, the force of the air becomes great enough to break them loose from the filter, allowing them to re-enter the air stream.)

This type of air cleaner does not use prefilters and mounts directly in the return air system.

Efficiency of this system is claimed to be 99% for pollens and spores (10-100 microns), 90% for dust (.01 - 30 microns), 90% for smudging (.01 - 5 microns) and 90% for tobacco smoke (.01 - 1.0 microns).

ELECTRONIC AIR CLEANERS

Originally, the EAC became popular because of its ability to remove pollen and spores which are the main cause of allergies and hay fever. Now, with added emphasis on pollutants of all types and a general awareness of the many other advantages of an EAC, it will surely continue to increase in sales and take a larger share of the market.

In the past, servicemen were reluctant to work on an EAC, mostly because they didn't fully understand its operation, but

FIG. 12-1.   Capability range of mechanical filter and electronic air cleaner. *Courtesy Honeywell, Inc.*

also because they were concerned that the high voltages that are generated would pose a shock hazard. Actually, even with the high voltage, the current draw is very low and the shock hazard is small. The operation of the EAC is fairly simple and the basic principles involved are covered in Chapter 1, Basic Electricity. To briefly review these principles:

- A molecule contains atoms with a positive nucleus and orbiting electrons with a negative charge. When the number of electrons equals the sum of the positive charge in the nucleus, the molecule is electrically neutral.
- Electrons can be displaced from the atom which results in free electrons (-) and an unbalanced molecule which becomes a positive ion.
- Like charges repel and opposite charges attract.
- DC current has constant polarity, that is, one side of the line is always positive and the other side is always negative.

OPERATION

An EAC has three main functional parts: a control system to operate the EAC, a power pack to provide the high voltage required (up to 10,000vdc), and an ionizing collector cell to precipitate the particles out of the air, Figure 12-2.

The ionizer and collecting cell are a complete assembly that slides out of the cabinet in one or two sections. It is normally reversible, so the EAC can be located almost anywhere in the furnace return air system and still maintain proper direction of the air flow. Clearance is required in the front to slide the cells in or out of the cabinet. The power pack and junction

box are mounted on the cabinet that holds the cell, with the control wires running to the furnace.

The system may also have a prefilter of aluminum mesh or regular filter media to trap the larger particles (10 microns or more) in the air stream prior to their entering the ionizing section. A complete functional layout is shown in Figure 12-3.

The ionizing section consists of a series of very-small-diameter tungsten wires that are connected to the 8000vdc output of the power pack. This high voltage creates an electrostatic field around each wire, that is positively charged. As the air (containing contaminant particles) passes through this field, free electrons (-) are created and are drawn toward the

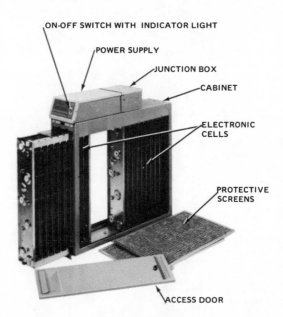

FIG. 12-2. Partially disassembled view of electronic air cleaner. *Courtesy Honeywell, Inc.*

wires with increasing velocity. They will collide with other air molecules, releasing more free electrons on the way. This produces positive ions (due to the loss of the free electrons). The positive ions attach themselves to the contaminant particles which assume a positive charge and move rapidly toward the grounded collector plates, which have a negative charge. Thus, some particles assume a positive charge and move to the grounded, or negative collector plates, while others have a negative charge and are attracted to the positive collector plates. The particles carried in the air stream will, therefore, be attracted to one or another of the plates in the collector section and will be held to that plate by electrostatic action and natural adhesion from oily contaminant particles like tobacco tar or cooking fat. Other forces are also acting on an individual particle, including resistance in the airstream, attraction or repulsion by other particles, gravity and inertia. However, most particles in the air (approximately 95% of those in the .01 to 10 micron size range) will be attracted to one plate or the other. A small amount of the particles will also be collected by the ionizer wires.

When these particles do collect on the plate, they lose their opposite charge and assume the charge of the plate. As the particles are removed from the air stream, they build up on the plates, insulating the plate and reducing its efficiency. The plates must therefore be cleaned periodically to maintain proper operation.

OZONE

An EAC will produce small amounts of ozone in normal operation. Ozone (O_3)

is an oxygen molecule with an extra oxygen atom and is a normal trace element in the atmosphere. It is created by solar radiation or lightning in the atmosphere and by arc welding or other electrical discharge on the earth's surface. High concentrations of ozone (defined as .05 parts per million, or more) can be irritating to the eyes and respiratory system.

Ozone readily oxidizes organic matter and is used widely to remove bacteria or algae, to purify water and to remove odors. Ozone will deteriorate rubber and some fabrics, so the EAC should always be run only when the furnace blower is on. Ozone does have a distinctive odor which can be more easily detected when other contaminants are removed, but this does not mean that the concentration is above normal safety limits. An EAC will produce only about .01 ppm of ozone, which is well below the allowable limit. The ozone concentration will be higher on a dry day than a humid day, which is another reason to limit the output well below the safety area of .05 ppm.

If the ozone level is higher than the homeowner desires, it can be reduced by lowering the voltage to the ionizer. On the Honeywell Model F50, this can be done by changing the leads to the transformer's resonator circuit, Figure 12-4.

APPLICATION

The EAC is a self-contained unit that is installed as part of the house return air system. Because the efficiency of an EAC decreases as the amount of air handled increases, it should be sized as closely as possible to the actual amount of air circulated in the house.

Electronic air cleaners are installed directly in the home duct system ahead of the furnace blower. Do not install downstream of the blower. If a prefilter is used, it installs ahead (upstream) of the EAC and the regular furnace filter is removed. The EAC is not position-sensitive, so the

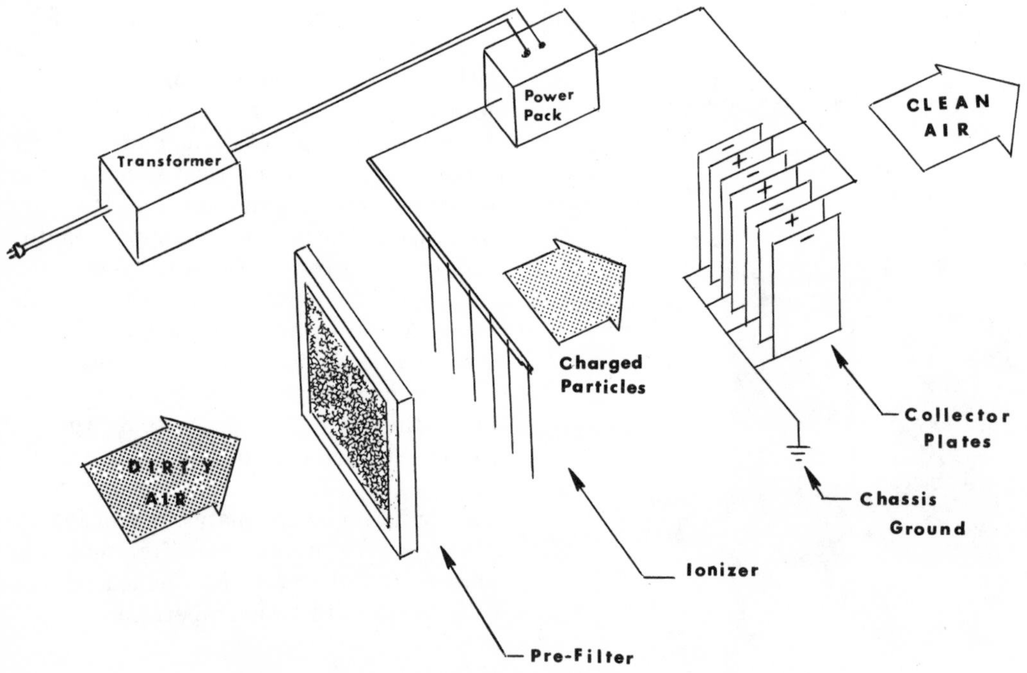

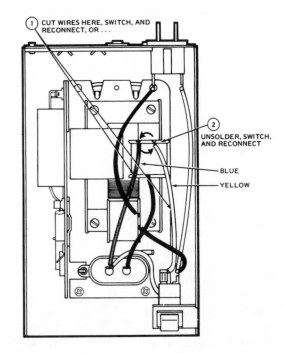

FIG. 12-3.　Functional view of electronic air cleaner.

FIG. 12-4.　Switching wires in the resonator circuit reduces voltage and ozone levels. *Courtesy Honeywell, Inc.*

same model may be installed in the system adjacent to an up-flo, down-flo, or horizontal furnace, Figure 12-5. When the duct has bends or angles close to the EAC, turning vanes, Figure 12-6, are recommended to balance the airflow.

There will be an arrow on the cell indicating the direction of air flow and the EAC must be installed with the air flowing in the right direction for proper operation.

Models are available that fit into the return air grille, but their use is limited because, in order to filter all the air being circulated, an EAC would have to be installed in every return air opening.

When an EAC is installed on a system that also contains a humidifier, care should be taken to make sure that no wa-

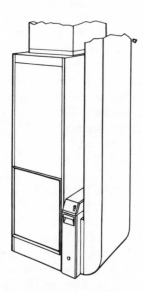

FIG. 12-5. Typical installation of an electronic air cleaner. *Courtesy Honeywell, Inc.*

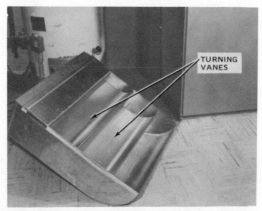

FIG. 12-6. Turning vanes. *Courtesy Honeywell, Inc.*

ter from the humidifier will reach the EAC. Water is a conductor and will cause popping and cracking, due to arcs between the plates. Also, the minerals in the water will deteriorate the plates, which greatly reduces the life and efficiency of the EAC. With an atomizing humidifier in the system, a standard furnace filter should be inserted between the humidifier and the EAC to reduce the buildup of mineral deposits.

When cold, outside air is introduced into the system, it should be warmed or premixed before it comes in contact with the cell. The return air temperature, across the cell, should not be allowed to drop below 40° F. Bringing the outside air in well ahead of the EAC, or using mixing baffles or vanes, will help. If a large amount of cold, outside air is introduced, it may have to be preheated.

Finally, the furnace and duct openings should be cut as close as possible to the dimensions of the EAC openings to avoid excessive pressure drop. Existing filter

rails can be removed and, when the EAC mounts directly to the furnace, the duct flange can be removed. In cases where the EAC is larger than the furnace opening, a transition must be used to hold pressure drop to a minimum and to avoid turbulance. A transition is almost always required on a down-flo application.

ELECTRICAL COMPONENTS

All EAC's will have essentially the same components, but the wiring and voltages produced may vary somewhat. Note should be taken immediately that all air cleaners will have both alternating (ac) and direct current (dc) and diagrams should be checked carefully to determine the type of current in every circuit. The following discussion will show which current is involved, by indicating vac or vdc after the voltage. The basic components and their functions are as follows:

Pilot light The Honeywell models have a pilot light wired into the transformer's resonator circuit which will be on whenever the unit is functioning properly. It will go off when power is disconnected, when the door is opened, when service is indicated, or when the cells need cleaning.

Ammeter The Emerson model uses an ammeter as a visual indication that the equipment is functioning properly and to tell when it needs maintenance or service. The ammeter is wired into the line (120vac) voltage circuit.

Off-on switch An off-on switch is required so that the EAC section can be deenergized during the drying cycle.

Power disconnect or interlock The interlock is a positive means of disconnecting power to the EAC whenever it is necessary to remove cells for cleaning or service. This safety feature avoids a possible shock hazard.

Step-up transformer The transformer converts the 120 vac primary to a much higher voltage (approximately 4000vac) in the secondary. An EAC requires this very high voltage to be effective.

Rectifier circuit The rectifier circuit consists of two selenium rectifiers (diodes) and two capacitors. (The collector plate acts as one capacitor in some EAC models.) The purpose of the rectifier is to convert the ac voltage input to dc voltage for the collector and ionizer. The capacitors are used in a *voltage doubler* circuit which approximately doubles the voltage in one leg of the circuit.

Bleeder resistor This resistor *bleeds off*, to ground, any excess voltage remaining when the unit is de-energized. The Honeywell unit performs this function by having the interlock switch *short* across the plates when the door is opened. This eliminates any shock hazard.

Collector plates The function of the collector plates has already been explained. Spacing will be from 3/16 in. to 1/4 in. Every other plate is connected to the positive side of the rectifier circuit and the alternate plates are connected to the ground.

Ionizer wires The function of the ionizer wires has also already been explained. These are very small diameter wires, mounted with tension springs and insulators that are connected to the positive side of the rectifier circuit.

Some models will have additional components such as:

Built-in water wash This system will wash the cells in place and requires a water supply and drain connection.

Dry-cycle timer The dry-cycle timer has a clock motor which can be set from 1 to 3 hours. The timer de-energizes the EAC circuit to allow adequate drying time after the cells have been washed.

Blower "on" switch This switch assures that the blower will stay on and not cycle with the fan control during the drying cycle.

FIELD WIRING

The EAC is completely prewired at the factory. The only field wiring is a 120v power supply from the furnace electrical system that is connected either directly into the EAC or to a makeup box on the EAC. Some furnaces will have a terminal block with provision for the EAC connections.

For efficient operation with minimum ozone production, the EAC should operate only when the furnace blower is on. There are two ways to accomplish this. One is to wire the EAC into the fan control circuit so it will only run when the fan control turns on the blower. This method will usually be the most convenient with a single speed motor.

With a two-speed blower, the EAC *on* time is controlled by a sail switch. A sail switch is mounted in the return air duct and is exactly what the name implies. It has a plate, or sail, which extends into the airstream and that is mechanically connected to a normally open switch. When the air pressure in the return air duct increases (blower on), it moves the sail and closes the contact bringing on the EAC, Figure 12-7.

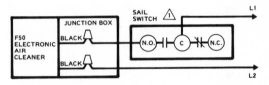

FIG. 12-7.   Schematic wiring diagram for sail switch. *Courtesy Honeywell, Inc.*

Note that *if* there is a blower relay installed in the system *and* the homeowner sets his thermostat for continuous blower operation, he will have adequate air for normal operation and have no problems on the drying cycle.

PERFORMANCE METER

Some EAC's, including the Emerson Slim Line model, are equipped with a performance meter. This is an ammeter, connected in the line-voltage circuit, that reads the current drawn by the unit. The meter face for the Emerson model is shown in Figure 12-8. The top illustration shows the needle position when the unit is ON and operating normally. If the needle fluctuates momentarily, around the ON position, this is of no significance. The center illustration indicates that the unit is OFF, with the needle all the way to the left. The needle will move to the right, into the SERVICE area, when the cell needs cleaning or if additional drying time is required, as shown in the lower illustration. The meter will also be in this position, if there are other problems requiring service, such as a broken wire or short in the system.

FIG. 12-8.   Emerson performance meter. *Courtesy Emerson Electric* Co.

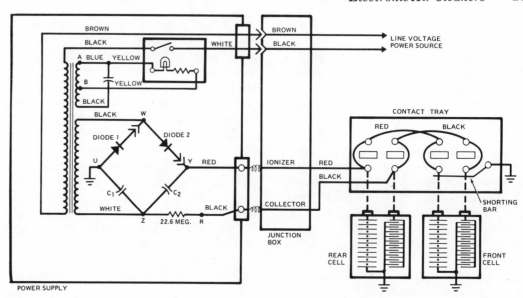

FIG. 12-9.   Schematic wiring diagram for electronic air cleaner. *Courtesy Honeywell, Inc.*

INTERNAL WIRING

The internal wiring and basic components of the power pack will be very similar for all EAC's. The ionizer voltage will be around 8000vdc and the collector voltage 4000vdc with respect to ground. A schematic for the Honeywell unit is shown in Figure 12-9.

All EAC's will have a step-up transformer to change the 120vac input into the secondary voltage of around 4000vac.

The Honeywell system includes a ferroresonant transformer which has a resonant winding, with a capacitor, in addition to the secondary winding, Figure 12-10. Voltage on the resonant winding is 150vac and its purpose is to maintain a relatively constant output voltage, even through the input voltage varies.

Regardless of load, the total output current is limited to 5 milliamperes which is safely below the shock hazard limit set by Underwriters Laboratories.

The power supply will not be damaged by a short across the output. If the load gets too large, the transformer drops out of resonance and both voltage and current decrease rapidly. This causes the pilot light, in the resonant circuit, to go out, indicating a service problem, Figure 12-9.

The rectifier circuit converts the ac voltage input to the dc voltage output required by the ionizer and collector sections. Note, too, that the 8000vdc voltage supplied to the ionizer section is approximately twice the 4000vdc voltage supplied to the collector section. For this reason, the rectifier circuit is often called a *doubler circuit*, Figure 12-11.

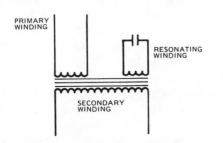

FIG. 12-10.   Schematic of ferroresonant transformer. *Courtesy Honeywell, Inc.*

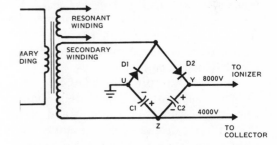

FIG. 12-11. Doubler circuit. *Courtesy Honeywell, Inc.*

A rectifier allows current to flow in only one direction. Thus, during the negative half of the ac waveform, current passes through D1 and the voltage (4000vdc) is stored in capacitor C1. During the positive half-cycle, current flows in the opposite direction through D2 and the voltage (4000vdc) is stored in C2. With the two capacitors in series, the output voltage to the ionizer is the sum of the voltages (8000vdc with respect to ground). The difference in voltage between the two sections gives efficient operation and helps prevent arcing.

MAINTENANCE

Basic maintenance of an EAC is the removal and cleaning of the cell at regular intervals. The frequency of cleaning will depend upon a number of variables. One is the construction of the house, as explained in Chapter 11. Loose construction will allow considerably more contaminant particles to enter the house than tight construction. If the occupants smoke or entertain a lot, the air cleaner will work harder and the dirt buildup on the plates will accelerate. A good rule of thumb is to clean the cells every 2 to 3 months, or oftener, if unusual conditions exist or the cells become very dirty between cleanings. The cells should be cleaned anytime the pilot light is out or the performance meter is in the SERVICE position.

One very simple method of cleaning the cell is suggested by Honeywell. Simply remove the cells from the unit and place them in an automatic dishwasher. Add dishwasher detergent and run the dishwasher through the complete washing /drying cycle.

Be careful not to damage the plates or break one of the ionizer wires when handling the cells. Keep in mind that the cell will be quite hot after the drying cycle, so allow it to cool off before handling. Be sure to drain any accumulated water from the tubes supporting the collector plates.

If a dishwasher is not available, the cell can be placed in any container large enough to hold it, such as a laundry tub or sink. The cell should be laid flat, so that it is completely immersed, and soaked in dishwasher detergent mixed with hot water. Calgonite is suitable and about four ounces per cell is required.

Several manufacturers offer a cleaning kit consisting of a large flat pan with a drain, detergent, and a hose with a spray nozzle. A garden hose will very satisfactorily wash the cell off after soaking. After soaking 20 to 30 minutes, slosh the cell around until it looks clean and drain the pan. Rinse both sides of the cell with warm water, while the pan is draining, by holding the nozzle at an angle to the cell. Stand the cell up, to remove as much water as possible, before replacing in the unit. Be sure to follow the arrows indicating the direction of air flow.

Wash and dry the screen, separately, *after* the cell has been cleaned. Lint from the screen can become lodged in the collector plates if the screen is washed at the same time as the cell.

After the cell is reinstalled in the unit, it must be thoroughly dried by running the furnace blower for at least an hour *with the EAC disconnected.* Units with a dry-cycle timer will automatically disconnect the cell until the drying cycle is completed and then re-energize the EAC pow-

er pack. Units without a dry-cycle timer are turned off by the manual switch, which must be turned on after the drying cycle is completed. During the drying cycle, the blower must run constantly so the fan control must be bypassed either by a switch on the unit or by setting the thermostat for continuous blower operation.

When the cell is removed from the unit for cleaning, certain other things should be checked.
1. Visually inspect the cell for any particles lodged in the plates which could cause a short.
2. Inspect for bent or shifted plates and straighten.
3. Inspect ionizing section for cracked or broken insulators.
4. Inspect ionizing section for broken wires. Remove pieces and replace. (The collecting cell will function, but at reduced efficiency, if a broken wire is removed and not replaced.)

SERVICE

Servicing an EAC is not difficult since most problems occur when the cells are dirty or have not been completely dried. The pilot light or performance meter will indicate when the unit requires attention. If washing and drying the cell does not correct the problem, then additional checks are indicated.

Tools required are: two screwdrivers with insulated handles, needlenose pliers, a test meter with a range of at least 12,-000vdc, soldering iron and neon test light (120v).

Most of the troubleshooting steps for the Honeywell F50 can be performed by observing the indicator light in the on/off switch. The light will be on whenever the transformer is working properly. A flow chart for electrical troubleshooting the Honeywell F50 is shown in Figure 12-12. With some variations, this basic procedure can be used with other brands of EAC, substituting ammeter readings or voltage readings for pilot light operation.

As was stated earlier, because of variations in voltages and test points, the service procedures suggested by the manufacturer of the EAC should be followed. Some general suggestions, following the sequence given in the chart, are as follows:

1. Be sure cells are clean and dry, and there are no broken wires or shorts.
2. Collector output can be checked by shorting across two plates. An arc or snapping sound indicates that the EAC is working properly.
3. Check the line voltage supply with a meter or test light to verify a 120v supply. If there is no supply, check backwards in the circuit through the switch to the power source.
4. Isolate the high voltage supply, disconnect the cells and check the output voltage. If correct, check the high voltage wiring. If incorrect, isolate the transformer and check the output voltage.
5. If the transformer output voltage is correct, check the rectifier circuit. If not correct, check the transformer windings for continuity.
6. Check the voltage across each capacitor in the rectifier circuit. If low, check the rectifiers.
7. A selenium rectifier can be checked by pushing on the end, at which the arrow points, with a pencil. If the spring will not compress, the rectifier is bad and

ELECTRICAL TROUBLESHOOTING

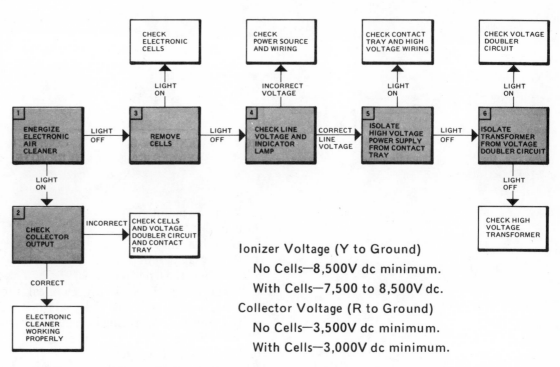

Ionizer Voltage (Y to Ground)
No Cells—8,500V dc minimum.
With Cells—7,500 to 8,500V dc.
Collector Voltage (R to Ground)
No Cells—3,500V dc minimum.
With Cells—3,000V dc minimum.

FIG. 12-12.   Troubleshooting flow chart. *Courtesy Honeywell, Inc.*

needs replacement.

8. Capacitors can be checked, *out of the circuit*, with a meter. A good capacitor will move part way on the scale and then to infinity. A shorted capacitor will show zero resistance.
9. On units with a bleeder resistor, turn on the unit for a few minutes, then turn it off. Wait about 10 seconds and then short across two adjacent collector plates with an insulated screwdriver. An arc indicates that the resistor is bad.

Detailed service procedures for the Honeywell F50 series can be found in their Manual 60-2067 A. Their manual on Application (70-9723) and Theory (70-9719) can be used to supplement this chapter on Air Cleaners. Emerson service manual SM-2 has similar information for their products.

Questions

1. What are pollutants?
2. How do they occur?
3. What is a particle?
4. What is a micron?
5. How big are visible particles?
6. What size particles will a mechanical filter remove?
7. What does an activated charcoal filter do?
8. Give the basic operating principles for an EAC.
9. What are the 3 basic parts in an EAC?
10. How does an EAC work?
11. What size particles will it remove?
12. What is ozone?
13. Is an EAC position sensitive? Why?
14. Give some application considerations for an EAC.
15. How is an EAC wired into the furnace?
16. What kinds of voltages occur at the ionizer?
17. What kinds of voltages occur at the collector?
18. What does a performance meter do?
19. How can you tell when the cell needs cleaning?
20. How often should the cell be cleaned?
21. Name four things to check for in regular maintenance.

Chapter 13
HEAT PUMPS

The heat pump or *reverse cycle air conditioning*, as it has sometimes been called, has been around for many years. The process of removing heat from the indoor air and getting rid of it to the outside air, through heat transfer using refrigerant, is standard procedure in air conditioning. Reversing this process, to provide heat indoors during the winter, was known to be possible and was widely promoted in the late 1940's and early 1950's. Unfortunately, the equipment used then was not properly designed for the added stress found in heat pump applications. Furthermore, installers and field servicemen knew very little about the heat pump and made problems worse.

As a result, the heat pumps of the early 50's failed at an amazing rate and the whole system fell into great disfavor. Most companies put their designs on the shelf and turned to other things. Fossil fuels were inexpensive and in good supply, electricity was cheap and the straight air conditioning market was growing rapidly so there was no incentive to promote an unreliable product.

Much of this has changed. With the oil embargo in 1973 and the rapidly increasing cost of all types of energy, more efficient use of energy is mandatory.

Also, gas and oil are becoming less available for new construction.

More and more companies turned their attention back to the heat pump and now the market is the fastest growing segment of environmental control. From 1974 to 1978, sales increased 3.5 times to 560,000 units. That's 21% of the unitary air conditioning market.

OTHER FACTORS

In addition to energy use, there are other important reasons why a heat pump has become a practical method of residential heating and cooling. Some of these factors did not exist even a few years ago.
1. Homes are better insulated so the cost of equipment has come down.
2. Special compressors and coils have been designed, specifically for this application, that are more dependable.
3. No vents or chimneys are required and there are no products of combustion to pollute the air.
4. Equipment, especially the indoor section, takes less space.
5. Controls and reversing valves are dependable.
6. Considerable engineering work has been done to explore and understand what is actually happening inside the system so overall design is much better.
7. Application is more precise.
8. More and more men are being trained to install and service heat pumps correctly.

FIG. 13-1. With the hood on, the outdoor half of a heat pump looks like the condensing unit of a straight air conditioner. Note the reversing valve at the junction of the three vapor tubes. *Courtesy Amana Refrigeration, Inc.*

WHAT IS A HEAT PUMP?

Before we get into the service and installation of a heat pump, we need to understand how a heat pump works and how it differs from a straight air conditioner or, more importantly, a furnace. Remember that a heat pump uses the compressor during the *heating* cycle as well as the cooling cycle. This can make the controls and their functions seem backwards, to one used to straight cooling, and cause some confusion.

A heat pump is not simply an air conditioner with the capability of reversing refrigerant flow. It is a precise, carefully engineered, matched-component system designed to perform a specific function — that is, to heat and cool with the same package of components. The outside unit, Figure 13-1, will look exactly the same as a conventional condensing unit. The inside unit, Figure 13-2, will be somewhat smaller than a furnace. There are several basic types, such as water-to-water, water-to-air, and air-to-air. We will deal only with air-to-air, the type most generally used in residential applications. As with air conditioning, the heat pump can also be a packaged system, Figure 13-3, having all the components in one cabinet, or a split system, with the outdoor unit remote from the indoor unit. The following descriptions and procedures will apply to both.

HEAT PUMP COILS

Heat pump coils are required to alternately function as evaporating and condensing coils. Therefore, they are designed and

FIG. 13-2. The indoor half of a heat pump is somewhat smaller than a furnace. *Courtesy Bryant Air Conditioning Co.*

FIG. 13-3. A heat pump may be a packaged system. *Courtesy Amana Refrigeration, Inc.*

sized differently. This fact must be remembered in case one needs replacement. Fin spacing, face area and circuitry generally change and the two coils must be matched to the system and to each other. Terminology also changes and they will be referred to as the indoor and outdoor coils rather than the evaporator and the condenser.

COOLING CYCLE

In the cooling mode, a heat pump operates exactly the same as a straight air conditioner. There are, however, some additional controls and components, such as check valves, a reversing valve, defrost relay and changeover relay, that are mostly nonoperative on cooling, but which can affect cooling performance and service. A refrigerant flow diagram, Figure 13-4, shows that the circuit is essentially the same as with air conditioning. The pressures, temperatures, enthalpy and state are also essentially the same (270 psig discharge and 80 psig suction). The air temperature leaving the indoor coil is likely to be 5 to 8° higher than with air conditioning due to the increased cfm.

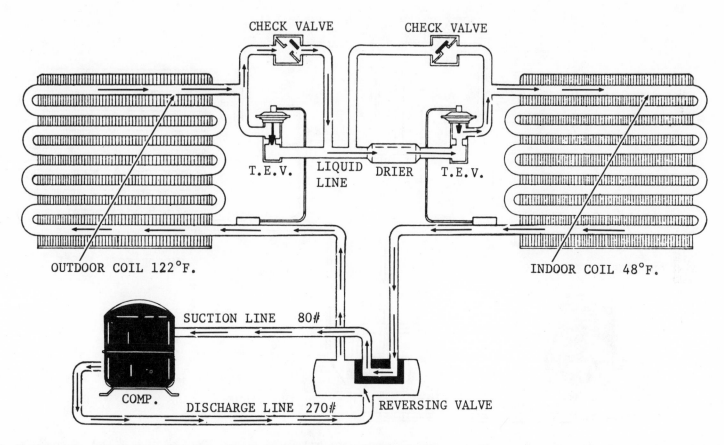

FIG. 13-4. Refrigerant flow while heat pump is in the cooling mode. *Courtesy Amana Refrigeration, Inc.*

HEATING CYCLE

When the system is changed to the heating mode, the function of both coils changes, Figure 13-5. The indoor coil now becomes the condenser and releases heat to the air stream in the house. The outside coil, as long as it remains at a lower temperature than the outside air, will pick up heat, which changes the liquid refrigerant to vapor. The vapor also picks up additional heat of compression from the compressor motor. Heat from both sources is delivered to the indoor air stream. As the indoor air absorbs this heat, the refrigerant is cooled and changes back to a liquid, releasing its latent heat. The liquid refrigerant is metered back to the outside coil to begin the cycle again. The refrigeration pressures will be less, 260 psig condensing and 38 psig evaporating at 30°F ambient, due to the fact that there is less heat in the air at 30° and the unit is doing less work. Both pressures will come down even more at lower ambients.

With a 115° condenser, the *temperature rise* over the indoor coil will be considerably less than from a conven-

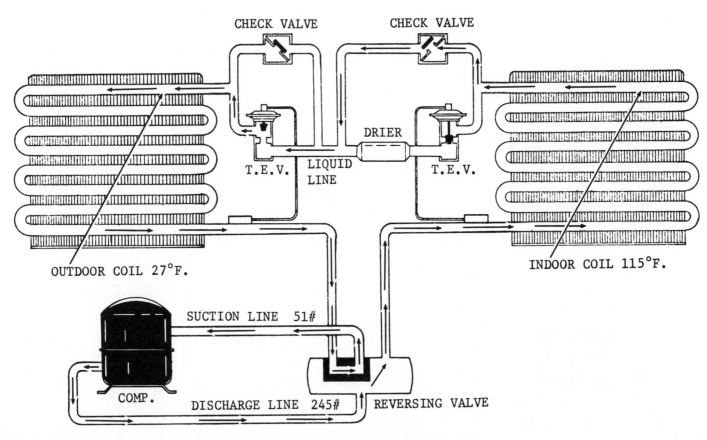

FIG. 13-5. Refrigerant flow while the heat pump is in the heating mode. *Courtesy Amana Refrigeration, Inc.*

tional furnace. The *delivered* air temperature to the room will be about 105°F, when it is 30°F outside, or only about a 35° rise over the coil. On the other hand, a gas or oil furnace will have an 80° rise and a delivered air temperature of 150°F. The air will *feel* even cooler, due to the increased volume of air, although the velocity will be about the same. This modest temperature rise will not affect the system's ability to maintain the temperature desired by the homeowner, but it will mean that the system will run for a longer period of time.

AIR FLOW

Air Conditioning & Refrigeration Institute (ARI), the industry standards association, has established that, for air conditioning, air flow should be 400 cfm per ton over the evaporator and 700 cfm over the condenser. However, since heat pump coils must function as either condenser or evaporator, the air flow over each must be equal. This has been set at 450 cfm/ton. This is a greater volume of air over the inside coil than with a straight air conditioning system. So, on a conversion replacement or add-on to an existing furnace, the duct system may be undersized. There is also much less air than normal flowing over the outside coil, so the heat rejection on air conditioning is less efficient while, on heating, the pick-up of heat is better. Some systems use a two-speed motor on the *outside* coil that will deliver more air in the cooling mode. This will hold down pressures and allow the unit to operate more efficiently on the cooling cycle when more air is needed over the outside coil.

RATINGS

Heat pumps are rated, as are single-purpose air conditioners, in terms of tons of cooling. However, this nominal rating also indicates the heating capacity of the unit, if you bear in mind that two temperatures are used to certify the rating.

One manufacturer, for instance, lists the ARI-certified rating of a typical 2½ ton heat pump as: 30,500 Btuh cooling capacity at 95°F outdoor temperature and 31,000 Btuh heating capacity at 47°F outside air temperature. At higher or lower ambient temperatures, the heating and cooling outputs diverge. Usually, the heating output at various ambients are given in 5° increments.

The selection of a heat pump for a given structure is made on the basis of its cooling capacity and matched to the calculated heat gain at a 95°F design temperature. It should not be oversized by more than ½ ton, or 6000 Btuh, even though additional capacity may be needed in the heating mode. As with air conditioning, control of latent heat is lost with an oversized unit, because of shorter running time, and undersizing will not handle the load.

Additional heating requirements are supplied by supplemental electric resistance heaters, in 5 to 10 kw increments. The number of electric heaters needed is determined by the calculated heat loss of the structure.

These electric resistance heaters are often called *strip* heaters because of their shape, Figure 13-6.

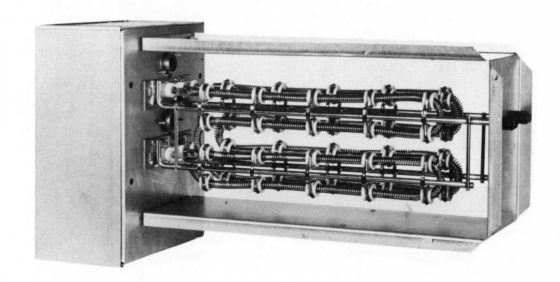

FIG. 13-6.   Electric resistance heater. *Courtesy Gould Inc.*

COEFFICIENT OF PERFORMANCE

The ratio of heat output (Btuh) to its power consumption (watts) is called *Coefficient of Performance* or COP for short.

Electric resistance heat is considered to be 100% efficient, having an energy output to energy input ratio of 1 to 1. That is, a kilowatt of electricity will produce 3412 Btu's of heat energy at any outdoor temperature. A heat pump, on the other hand, can improve on this energy efficiency. The amount of heat energy provided will depend upon the outdoor temperature, but it will always be greater than 1 to 1, even down to a temperature of zero and below.

Contrary to what many people may believe, there is always a certain amount of heat in the outside air, even at 0°F. Air does not lose all of its heat content until it reaches *absolute zero*, which is -460°F. As the temperature goes down, it becomes more difficult to extract the heat from the air, but it is still available and a heat pump can use this heat with more efficient consumption of electric power than resistance heat.

For example, one manufacturer lists the following output data for one of its heat pumps:

TABLE 13-1

Ambient Air Temp.	Btuh Output	Watts Input	Electric Heat Output	COP
60°	50,000	4400	15,012	3.33
40°	38,000	3400	11,600	3.27
20°	26,000	2700	9,212	2.82
0°	16,000	2100	7,165	2.23

This shows that with an electrical input of 4400 watts at 60°F outside air temperature, the heat pump will produce 50,000 Btu's of heat energy for use in heating a home. An electrical resistance heater, with the same 4400 watts input, will produce only 15,012 Btu's (4.4 x 3412). Dividing 50,000 by 15,012 gives a COP of 3.33, making the heat pump better than *3 times* as efficient as a resistance heat unit at 60°F. Even at 0°F, the heat pump is twice as efficient in producing usable heat from a given amount of electricity.

ENERGY EFFICIENCY RATIO

Another commonly used measure of the efficiency of air conditioners, heat pumps and appliances is the Energy Efficiency Ratio (EER). EER is the ratio of the energy produced directly, in Btu's, to the energy consumed, in watts. It gives the user an idea of how well the appliance used the energy it consumes and what the relative operating cost might be. The higher the number, the more efficient the product. In the example of COP, at 60°F the unit produced 50,000 Btu's of heat with an input of 4400 watts. Dividing 50,000 by 4400 gives an EER of 11.36 and at 0°F an EER of 7.6 (16,000 ÷ 2,100). This is an overall average of about 9.5, which is very good. For cooling, this same unit would produce 31,000 Btu's at 95° db/67° wb (ARI rating standard) and will use 3600 watts for an EER of 8.6. This is also a very good EER, since the basic code standard, being developed by many cities, is 8.0.

BALANCE POINT

A *balance point* of a particular house is shown in Figure 13-7. To construct this chart and select equipment, certain facts must be known.
1. Cooling load — assume 48,000 Btu
2. Heating load — assume 90,000 Btu
3. Design temperature — assume 0°F
4. Heat pump capacity at 2 ambient temperatures, given by manufacturer in his specs — assume 46,000 at 45°F and 26,000 at 20°F

For this type of calculation, the industry has established 65°F as the point where neither heating nor cooling is required.

The chart shows the outdoor air temperature, in 5°F increments, along the horizontal scale and the building heat loss, in 5,000 Btu increments, on the vertical scale. The first point (A) is at 90,000 Btu/0°F, which is the heat loss at design conditions. The second point (B) is at 0 Btu/65°F, where no heat is needed. Line 1, connecting points A and B, gives the heat loss at any ambient.

Line 2 will plot system capacity. The first point (C) is given by the manufacturer as 46,000 Btu/45°F. Point D is given as 26,000 Btu/20°F. Line 2, connecting points C and D, gives the system capacity for the heat pump at any ambient.

Line 2 shows that heat pump capacity declines rather sharply as the outdoor temperature falls, but even at 0°F it is producing 10,000 Btu's of heat. Where the two lines cross (E) is the point where the heat pump alone is exactly matched to the heat loss in the house. This is Balance Point 1 (BP1). Any additional drop in outside temperature will create a need for supplementary heat.

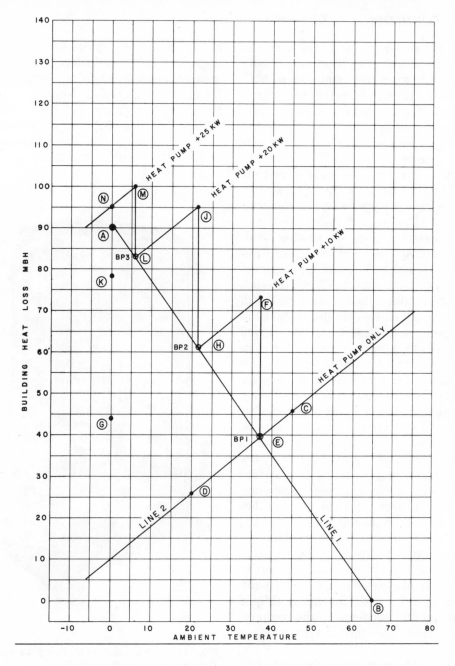

FIG. 13-7. Typical balance point chart. *Courtesy Rheem Air Conditioning Div., City Investing Co.*

Assume the addition of a 10 kw resistance heater. This will add 34,120 Btu's of system heating capacity. Plot 34,120 Btu's up from BP1 (heat pump alone) to point F. Then, plot 34,120 Btu's up from the 0°F point on Line 2 (heat pump at design temperature) to point G. Draw a line between points F and G until it intersects Line 1 at point H. This is the second balance point (BP2), using one 10 kw heater plus the heat pump, and falls at about 22°F outside temperature where the heat loss is 62,000 Btu's.

Since the system still cannot handle the 90,000 Btuh heat loss, another 10 kw heater and an additional 34,120 Btu's must be added. Repeating the procedure used to establish BP2, locate points J and K and Balance Point 3 at L. The system, with two 10 kw heaters, can now go down to about 6°F or 83,000 Btu's.

This is still short of the design requirements, so another heater must be added. However, the last stage can be 5 kw rather than 10 kw. Plotting points N and M from BP3, shows that the system, with 25 kw of resistance heat plus the heat pump, can produce 95,000 Btu's of heat at the design temperature of 0°F. This will be more than adequate.

This system also meets a requirement of many city codes that states that the *emergency* or additive heat be 80-90% of the total system capacity at design conditions. This is to assure at least minimal heat if the compressor should fail. This example will produce 94% (85,300 ÷ 90,000).

NOTE: The resistance heaters are added downstream, beyond the coil. This allows the air to be heated first by the refrigeration system, as much as possible. The minimum amount of resistance heat is then added, to satisfy the thermostat.

HEAT PUMP COMPONENTS

There are several components added to a heat pump system that are not found on a straight air conditioning system.

Most of these added components function on the heating cycle only, since the cooling cycle is exactly the same as with a straight air conditioning system.

COMPRESSORS

The single-purpose air conditioner is rated at a 95°F ambient and, therefore, is a high temperature machine. In a heat pump, this same air conditioning compressor may be operating in 0°F ambient air and be expected to perform as a low temperature machine. It could do both, if the density of the refrigerant vapor remained constant. However, R-22 vapor at 0°F is only 18% as dense as vapor at 100°F and comparably fewer pounds of refrigerant can be pumped by the compressor.

Refrigerant-22 is almost universally used in heat pumps. It has excellent characteristics for cooling but is very poor at low ambients, because of its low density and specific heat. However, it is widely available and less expensive than R-502, which has better temperature qualities, but higher operating pressures.

The compressor's performance at low ambients can be approximated from Table 13-1. Obviously, the heat output at 60°F is over three times the output at 0°F, largely due to density of the refrigerant.

Since the system was charged on the assumption of a 95°F ambient, it is grossly overcharged when the ambient is 0°F. With fewer pounds of refrigerant being vaporized in the outside coil, this excess refrigerant accumulates in the inside coil. When the cycle is reversed, during defrosting, this excess liquid may flood back into the compressor, cause oil foaming and the loss of lubrication. In extreme cases, this excess refrigerant may slug the compressor.

Compressor cooling is another challenge. To grasp the problem, reread Electric Heat Output, in Table 13-1, as the heat produced by the motor. At 60°F, the 50,000 Btuh total output is the sum of 35,000 Btuh taken from the outside air and 15,000 Btuh from the motor. Experience has proved that this is an adequate margin for compressor cooling. At 0°F, however, 9000 Btuh are extracted from the outside air and 7,000 from the heat of the motor. And, remember, the motor's heat is added after the limited amount of vapor has entered the compressor. Since there is such a modest amount of vapor available at low temperatures, motor cooling is a serious problem. To protect the motor windings, some manufacturers introduce a calculated amount of liquid refrigerant into the compressor to assure adequate cooling.

Intermittent start-ups in cold weather present other problems. Cold, congealed compressor oil increases the starting loads, and ampere draw, on the motor. Liquid refrigerant will condense in the compressor crankcase and threaten to foam on start-up.

Major compressor manufacturers now have special compressors for heat pumps, Figure 13-8. These include features like greater bearing surfaces, better motor insulation, immersion heaters, internal line break overloads and more efficient motors. Because of these special features, only compressors that are designated as heat pump models should be used for replacement.

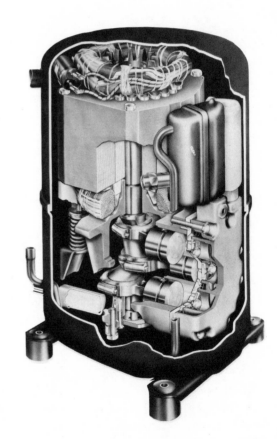

FIG. 13-8. Hermetic heat pump compressor. *Courtesy Tecumseh Products Co.*

CRANKCASE HEATERS

Crankcase heaters supply heat to the compressor crankcase at all times and *all* heat pumps should have them. This is because the heat pump is operating at low ambients and will have some *off* time. The heaters reduce the amount of liquid refrigerant in the compressor crankcase and maintain a reasonable oil viscosity.

LOW AMBIENT LOCKOUT

Some heat pumps will have a low ambient lockout. This may be just an outdoor thermostat, that shuts off the compressor at low ambients. The lockout point varies and can be as high as 20°F or down to 0°F or below. Some manufacturers advocate running the compressor at all ambients, so the manufacturer's specs should be checked.

Running the compressor constantly makes sense from two standpoints. One is that motor heat is always available, no matter what the ambient. An equally important factor is that, after shutdown in a low ambient, the motor must restart after the oil has thickened and there is a good chance of liquid refrigerant being in the crankcase which can slug the compressor.

LOSS OF CHARGE

On cooling, it is fairly easy to diagnose a loss of charge because the customer will experience a loss of cooling capacity. On heating, however, if the compressor goes out, the strip heaters will take over and heat the space, masking the problem for some time. The homeowner may not know anything is wrong until the electric bill arrives.

The internal overload may not be able to protect against loss of charge because the motor is not heavily loaded on heating at low ambients. The system will develop high discharge temperatures and corresponding high compression ratios that are damaging to the compressor although these factors may not be readily apparent.

It is normal for the compressor to operate at low suction pressures during low ambient conditions or when the outdoor coil is heavily frosted, so a low pressure control would not be an effective safeguard against loss of charge, since its setting would have to be much too low.

One device used to overcome this problem is a discharge line thermostat set at about 275°F, with a 150°F differential. It must be well insulated from the air flow, but normally will take the compressor off the line when head pressures go up due to loss of charge. Another is a low pressure control located in the liquid line, with a cutout set point of about 5 psig.

METERING DEVICES

As in air conditioning, the refrigerant flow can be controlled by either an expansion valve or capillary tube. At least one compressor manufacturer strongly advocates a capillary system for reasons other than first cost. This manufacturer feels that the only way the compressor can survive in low ambient conditions is by *controlled* flooding of the compressor, with liquid refrigerant, to cool the motor without affecting lubrication. A capillary tube is primarily a constant feed device and, on a heat pump application, operates at 10 to 15°F of superheat at a standard running

ambient of 47° db/43° wb. As the outdoor ambient drops, the superheat reduces and liquid flooding occurs at about 10°F ambient.

This is not as drastic a situation as most people would assume. Most compressors will tolerate a certain amount of liquid — how much depends upon the particular design. The liquid refrigerant, in passing over the motor and cooling it, will pick up heat and change into a vapor. In a sense, this is a secondary evaporation process and occurs before the refrigerant reaches the valves.

On the other hand, an expansion valve operates on a more or less constant superheat. When the evaporating temperature falls, the discharge temperature will increase which can cause oil breakdown. A low ambient lockout can prevent this, but can produce excessive operating costs.

Another problem is oil return to the compressor. When the evaporating temperature falls, the compressor capacity drops and the velocity through the entire system decreases. Also, the oil gets thicker and harder to move through the system. With a capillary system and liquid floodback, the oil is more easily moved back to the compressor because the oil readily mixes with the refrigerant.

Another factor is that there is much less refrigerant evaporated during the heating cycle than during the cooling cycle. The excess charge backs up in the indoor coil, because the expansion valve will only feed the amount of refrigerant that the outdoor coil can handle. On defrost, all of this liquid rushes back to the compressor, which can cause several problems such as broken valves or blowing the oil out of the compressor crankcase.

The expansion valve itself, because it has mechanical parts and a diaphragm, can be damaged from the reversing action of the system on initiation and termination of the defrost cycle.

CHECK VALVES

Since most heat pumps have two thermostatic expansion valves — one for each coil — the inoperative TXV must be bypassed to assure liquid flow to the unit that is metering. This is accomplished with a pair of one-way check valves, Figures 13-4 and 13-5. In the cooling mode, one check valve bypasses liquid around the outdoor TXV while the other check valve forces liquid flow to the indoor TXV. Conversely, in the heating mode, one check valve bypasses liquid around the indoor TXV, while the other check valve forces liquid flow to the outdoor TXV.

ACCUMULATORS

It has been pointed out that a heat pump on low ambients must cool the compressor motor with liquid refrigerant. It is also operating with low velocities which can cause problems in returning oil to the compressor. Most particularly when it shifts to defrost, a large surge of liquid is returned to the compressor. In order to minimize liquid slugging under these conditions, a suction line accumulator, Figure 13-9, is added to act as a receiver for the excess liquid and return it to the compressor in vapor form according to the pumping rate of the compressor at the time. It also acts as a storage point for refrigerant

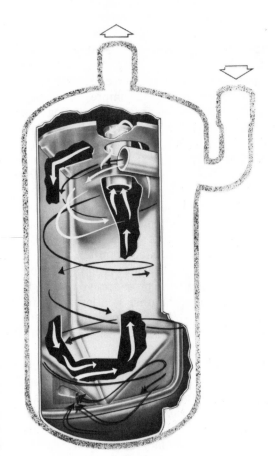

FIG. 13-9. Suction line accumulator. *Courtesy Tecumseh Products Co.*

during the heating cycle when much less refrigerant is required by the system.

REVERSING VALVES

Changing over the heat pump, from heating to cooling and vice-versa, is accomplished with a reversing valve, also called a *four-way* valve, Figure 13-10. The main valve body encloses a slide that moves from one end to the other and determines the direction of refrigerant flow. The center port, at the lower part of the valve, is connected to the compressor suction line. The port on the opposite side opens the slide chamber to the compressor discharge line. The two ports flanking the compressor suction port are connected to the vapor lines to the indoor and outdoor coils.

When the slide is in the heating mode, the vapor line from the outside coil feeds

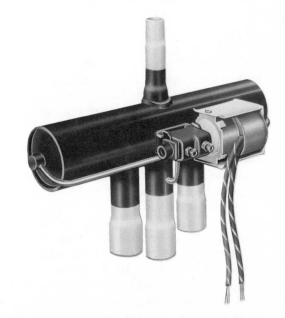

FIG. 13-10. Reversing valve. Three suction lines are clustered below, discharge line is above. *Courtesy Ranco Controls Div.*

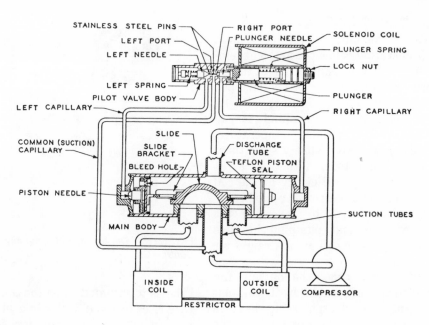

the main valve body, through capillary tubes linking the cylinders with the pilot valve. A third capillary tube, from the suction line to the main valve, is connected to the chamber between the needle valves. The suction capillary acts as a bleeder line for either of the capillaries and the related cylinders.

When energized, the solenoid opens the right port of the pilot valve and allows vapor in the right cylinder of the main valve to bleed off through the suction capillary which is never closed. The slide will now be drawn to the right because the pressure in the right cylinder is equal to suction pressure while the right piston is subjected to higher pressure from the discharge vapor in the center chamber. When it completes its travel, the piston seats a

FIG. 13-11. Cutaway view of reversing valve. *Courtesy Ranco Controls Div.*

the suction side of the compressor and the discharge line from the compressor is open to the inside coil. When the valve reverses, in the cooling mode or defrost phase, the vapor line from the inside coil is open to the compressor suction line and the discharge line is open to the outside coil, Figure 13-11.

Thus, the coil ports are alternately suction or discharge ports, while the vapor flow through the compressor is in the same direction at all times.

The position of the slide in the main valve body is controlled by a pilot valve that is operated by a solenoid, Figure 13-12. The pilot valve has two spring-loaded, nylon needle valves that control the pressure in the cylinders found at each end of

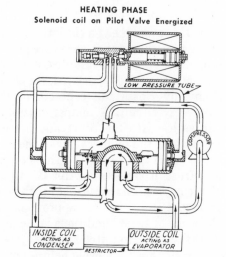

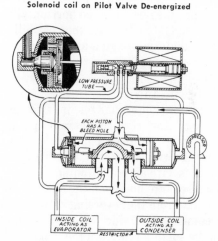

FIG. 13-12. Refrigerant flow through reversing valve during heating (left) and cooling (right). *Courtesy Ranco Controls Div.*

nylon needle, formed in the head of the piston, and seals the capillary tube to the right cylinder.

When the solenoid is deenergized, its spring will close the right needle and push open the left needle, using a pin that links the two needles. This will bleed the pressure in the left cylinder and allow discharge pressure on the left piston to move the slide to the left. Again, the capillary tube is sealed when the piston completes its travel. With the slide valve at rest, pressures in all parts of the main valve body are equalized through bleed holes in the pistons.

The reversing valve solenoid can have either a 24v or a 240v coil that can be energized on either the cooling or heating cycle, depending upon the type of thermostat and coil piping arrangement specified by the manufacturer.

Reversing valves are not generally position sensitive, but most manufacturers suggest that they be mounted horizontally, where possible, to eliminate the possibility of foreign material lodging at one end and causing improper seating.

CONNECTING LINES

There are some changes in size and terminology for the refrigerant piping used in heat pumps. The liquid lines between the two coils will be about the same size as with air conditioning.

The discharge line from the outside coil and the line from the reversing valve to the compressor suction side will be the same size as the suction line from the inside coil. In a heat pump, these large diameter pipes are called vapor lines.

DEFROST CONTROLS

Because the outdoor coil of a heat pump is operating below the ambient temperature, water will condense out of the air and form frost on the coil. This will happen most frequently on humid days with an outdoor temperature of around 45° to 35°F. Below 20°F outside temperature, frost accumulates very slowly since there is very little moisture in the air. Frost buildup reduces the efficiency of the coil and the coil must be defrosted periodically. This is accomplished by reversing the refrigerant flow and pumping hot gas through the outside coil as in straight air conditioning.

At the same time, the outside fan motor is turned off, so the frost will melt faster, and the electric strip heat is brought on, so the delivered air to the house will be warm.

The most common defrost control system is a combination time/temperature system. This system has a clock timer that runs any time the compressor is operating. Approximately every 90 minutes, it can be set for shorter periods if desired, the timer mechanically closes the contacts in the defrost relay circuit. A temperature sensor, in series with the timer and defrost relay, closes if the temperature of the outdoor coil drops to about 32°F. Thus, to initiate defrost, it is necessary for the timer to have completed its cycle *and* for the coil temperature to drop to approximately 32°F. If one of these conditions is lacking, no defrost cycle is initiated and the system waits for another time cycle before trying to defrost.

If the system does go into defrost, the

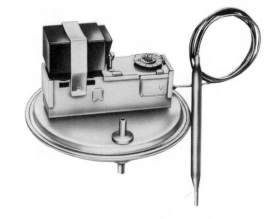

FIG. 13-13. Defrost control. *Courtesy Robertshaw Controls Co.*

cycle is terminated when the coil temperature rises to about 65°F. On a 40° ambient day (when ice accumulation is usually heaviest) this may take 5-6 minutes. At lower ambients, say 10°F, the cycle may take only three minutes. In most systems, the timer can override the thermostat and will terminate defrost after ten minutes even if the coil does not reach 65°F because of high winds or very low ambient.

Another type of defrost control operates on pressure initiation/temperature termination, Figure 13-13.

Sensors 1 and 2, Figure 13-14, monitor air pressure at the inlet and outlet of the coil. On the *push-through* unit shown, sensor 1 is needed to counteract any wind condition while sensor 2 checks the increase in pressure as ice builds up. Typical pressure readings, in inches of water column, will be 0.2" for a clean coil, 0.5" for a coil with 50 to 75% ice coverage, and 0.65" for a coil that is completely blocked. The control can

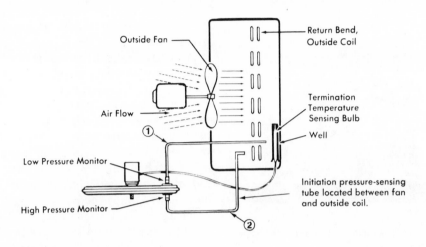

FIG. 13-14. Defrost control measures air pressure drop across the coil. *Courtesy Robertshaw Controls Co.*

be set, using an adjusting screw, to initiate defrost anywhere within a range of 0.15" to 1.0" w.c. There is no timer involved, so the sensor initiates defrost only when needed.

The termination temperature sensing bulb is located in a well in the leaving air-stream from the coil. It can be set to terminate defrost at any point within a 50 to 90° range and is usually set at 65°F.

In both systems, the defrost control operates through a defrost relay, usually DPDT, that energizes the changeover relay, turns off the outdoor fan motor, and energizes the strip heat. The coil of the defrost relay is usually 240v, even though it looks identical to the changeover relay which is 24v. If these coils are removed for any reason, be sure they are properly identified, as they are *not interchangeable!*

NOTE: Under certain temperature and humidity conditions, a rather large vapor cloud can be formed during defrost. The homeowner and serviceman should be aware of this and not become alarmed. *It is normal* and shows the defrost system is functioning properly.

INSTALLATION

Certain service, piping, wiring and air flow clearances are indicated in the manufacturer's installation instructions and these must be followed for a good installation. These will be similar to an air conditioning installation.

Heat pumps will have drain holes or slots in the cabinet to allow the water formed during defrost to drain out. Since defrost occurs in the winter, these holes must be free of any obstruction, to prevent ice from forming inside the unit. They should be checked every month and after every heavy snow or ice storm.

Normal practice is to use a mounting pad or mounting frame to raise the outdoor unit at least 6" off the ground. In areas where heavy snowfall is likely, the mounting height is increased to 20 or 24". This can be done by mounting the unit on cement blocks or by building an angle iron frame for it, Figure 13-15. A gravel apron is also suggested to aid in water drainage.

The unit should be located in a reasonably open area where drifting snow will not pile up around it.

The prevailing winds should not blow directly through the coil section (difficult with round or 3-sided units). There should be enough clearance from the building so water or ice from the roof cannot drop directly on it or in front of the coil.

Proper mounting and installation must allow the defrost water to drain away and not refreeze on the unit.

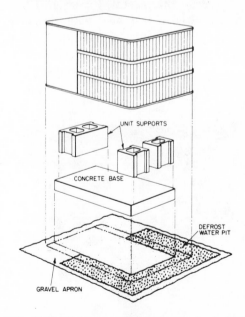

FIG. 13-15. Recommended procedure for installing a heat pump condensing unit.

ELECTRICAL CIRCUITRY

Before going into heat pump circuitry, review the basic wiring and switching functions of the special thermostats used with heat pumps. In the diagrams that follow, the internal switching of the thermostat will not be shown in order to devote more space to the overall circuitry.

Note that a heat pump will have three very distinct circuits, interconnected primarily by the various controls which can be either line voltage or low voltage and in some cases both. Separate disconnects are required for the 240v compressor and electric heat circuits. It is also possible that an additional disconnect will be required for the heaters due to the load.

THERMOSTATS

A special thermostat, Figure 13-16, is normally used with a heat pump because of the additional functions required. This

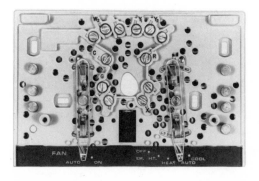

FIG. 13-17. Subbase for a heat pump thermostat. *Courtesy Honeywell, Inc.*

can be either a three-bulb or a four-bulb thermostat.

As with air conditioning, a sub-base is needed for the various switching functions, Figure 13-17. A heat pump thermostat will often include an *emergency heat* switch to energize the electric strip heaters in case of compressor failure or while the compressor is shut down for service. This is a manual switch that bypasses the other controls. There will also be an indicator light that reminds the homeowner that he is heating with electricity alone.

These thermostats will have two levers to set the desired room temperature, one

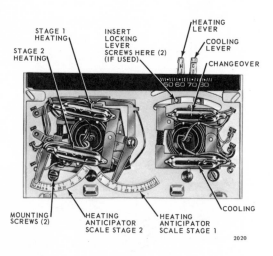

FIG. 13-16. Four-bulb thermostat. *Courtesy Honeywell, Inc.*

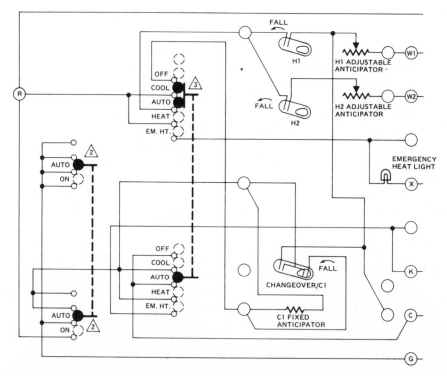

FIG. 13-18. Schematic wiring diagram of a three-bulb thermostat. *Courtesy Honeywell, Inc.*

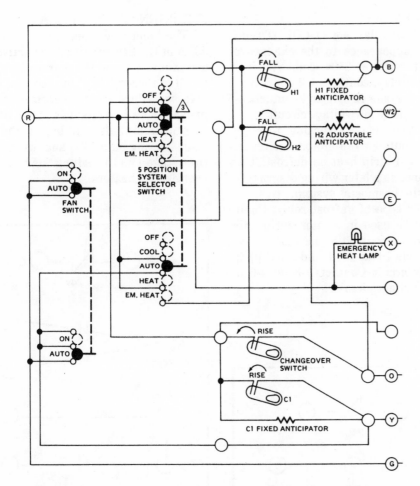

FIG. 13-19. Schematic wiring diagram of a four-bulb thermostat. *Courtesy Honeywell, Inc.*

heating bulb energizes the reversing valve, shifting the refrigeration system over to heating. The second-stage heating bulb brings on the supplementary electric heat when the bulb *makes*, usually at about 2° below the first stage setting. Thus, with a three-bulb thermostat, the reversing valve is normally energized during the *heating* cycle.

A four-bulb thermostat, Figure 13-19, uses the first-stage cooling bulb to energize the reversing valve and the second-stage cooling bulb to bring on the compressor when there is a call for cooling. On heating, the reversing valve is de-energized and the first-stage heating bulb brings on the compressor. If the compressor alone cannot maintain the temperature, the second-stage heating bulb will bring on the strip heat at about 2° below the first-stage setting. Therefore, a four-bulb thermostat energizes the reversing valve during the *cooling* cycle.

If the heat loss of the structure exceeds the capacity of the compressor plus one stage of strip heat, then additional heaters are required at each new balance point,

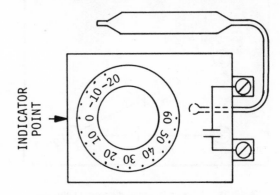

FIG. 13-20. Outdoor thermostat.

marked *heat* and the other marked *cool.* With the switch in *auto,* the thermostat will bring on either heating or cooling, at the preset room temperature levels.

Both types of thermostats will have fixed cooling anticipators. They will have adjustable heat anticipators applied to first stage only, both stages, or second stage only. The value will be about .4 amps

for the first stage and .5 amps on the second stage, depending upon the amount of strip heat used and the method of sequencing. These values should be checked.

A three-bulb thermostat, Figure 13-18, uses one bulb to control cooling, just as in regular air conditioning, and two other bulbs to control heating. Thus, heating can be energized in two stages. The first-stage

Figure 13-7, beyond that of the heat pump and one strip heater, taken together. The common way of activating these heaters is with outdoor thermostats, Figure 13-20, set at the auxiliary heater balance points.

CHANGEOVER RELAY

The changeover or reversing relay is a DPDT relay with a 24v coil, Figure 13-21. Contacts 1-2 and 1-3 are in the line voltage circuit and contacts 4-5 and 4-6 are in the low voltage circuit.

DEFROST RELAY

The defrost relay is also a DPDT relay, similar in appearance to the changeover relay, but with a 240v coil. As in a changeover relay, contacts 1-2 and 1-3 are in the line voltage circuit and contacts 4-5 and 4-6 are in the low voltage circuit.

These two relays, working together, control the outdoor fan motor, the reversing valve and strip heat on defrost. They will be shown in a later wiring diagram for a three-bulb thermostat system.

Another type of defrost relay, Figure 13-22, will be used in a later wiring diagram using a four-bulb thermostat system. Contacts 7-8 and 7-9, and the coil, are in the 240v circuit. Contacts 1-3 and 4-6 are in the low voltage circuit.

EMERGENCY HEAT RELAY

The emergency heat relay, Figure 13-23, is in the 24v circuit and is activated by a manual switch at the thermostat to bring on all of the strip heat in the event of compressor failure. It de-energizes the compressor and bypasses the outdoor thermostats. It also will bypass the second stage heating bulb (W_2) and, as long as the first stage (W_1) is calling for heat, it will energize the heaters.

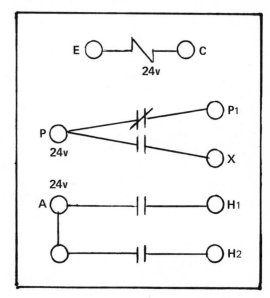

FIG. 13-23. Schematic wiring diagram of an emergency heat relay.

3-BULB COOLING

On cooling, heat pump operation is about the same as a single-purpose air conditioner. With the three-bulb thermostat set on cooling, the reversing valve is de-energized, the outdoor coil is the con-

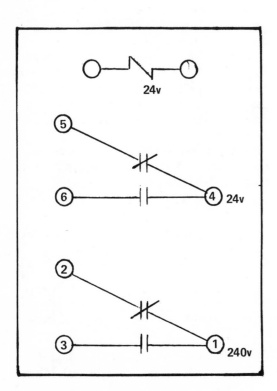

FIG. 13-21. Schematic wiring diagram of change-over relay.

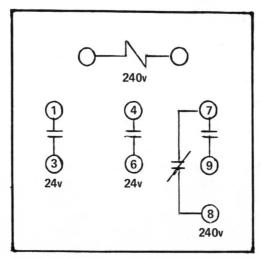

FIG. 13-22. Schematic wiring diagram of a defrost relay.

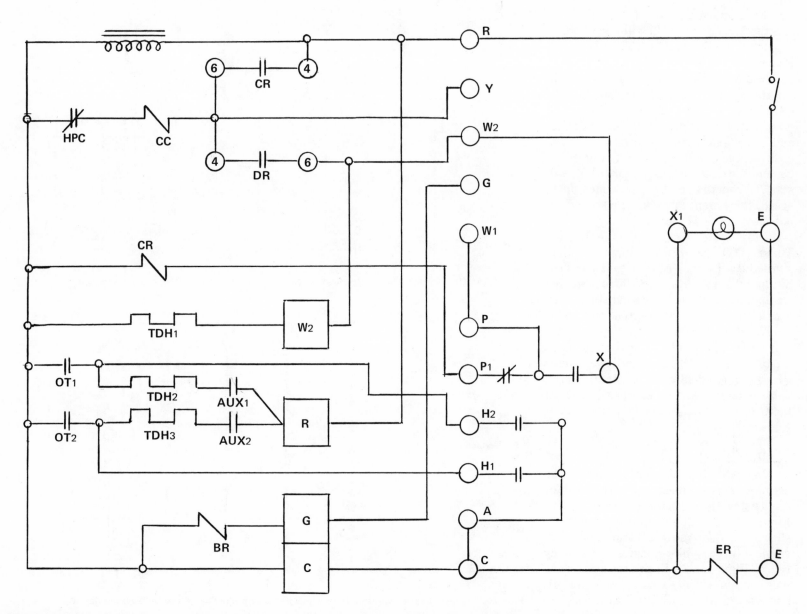

FIG. 13-24. Schematic wiring diagram showing the low voltage circuits of a heat pump connected to a three-bulb thermostat.

denser, and the indoor coil is the evaporator.

The Crankcase Heater (CH) will be on and the Transformer (T) will be energized whenever the line voltage switch is closed, Figure 13-25.

On a call for cooling, the Y terminal in the thermostat is energized, Figure 13-24, completing a circuit to the Contactor Coil (CC). If the High Pressure Control (HPC) and other safety devices are in their NC position, the Compressor (CSR), Figure 13-25, and the Outdoor Fan Motor (OFM) will start. When the compressor comes up to speed, the Potential Relay (PR) will drop out the Start Capacitor (SC).

Many systems will run the Defrost Timer (DT) motor even during cooling, since there is little chance that the heat pump will go into the defrost mode. Ordinarily, the coil temperature will exceed the setting of the defrost sensor.

The Blower Relay (BR), Figure 13-24, will be energized when the thermostat is set to *on*, where the Indoor Blower Motor (IBM) Figure 13-26, runs constantly, or on *auto*, where it will be energized through the G terminal and come on with the compressor.

3-BULB HEATING

On heating, operation of the heat pump becomes more complex, due to the need for sequencing the Resistance Heaters (RH), Figure 13-26. The sequencing system shown in Figures 13-24 and 13-26 will employ three White-Rodgers time delay relays, Figure 13-27. Each time delay relay will contain a Time Delay Heater (TDH) and a set of NO Auxiliary Contacts (AUX)

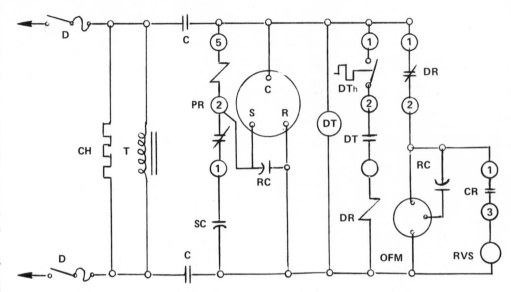

FIG. 13-25.   Outdoor line voltage circuits for a heat pump controlled by a three-bulb thermostat.

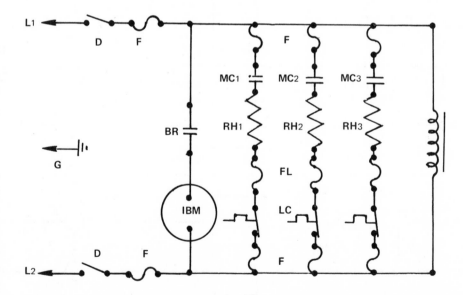

FIG. 13-26.   Indoor line voltage circuits for a heat pump controlled by a three-bulb thermostat.

FIG. 13-27. Time delay relay. *Courtesy White Rodgers Div., Emerson Electric Co.*

in the low voltage circuit, plus a set of NO Main Contacts (MC) in the line voltage circuit. Two of the three time delay relays will have NO Outdoor Thermostats (OT) in series with the Time Delay Heaters (TDH).

With the thermostat set on heating, a fall in temperature will complete a circuit through the first-stage heating bulb and terminal W_1. The circuit is completed, through the NC contacts, P and P_1 of the Emergency Heat Relay, to the coil of the Changeover Relay (CR) and back to common (C). Energizing the Changeover Relay coil will close NO Changeover Relay (CR) contacts 4 and 6 in the low voltage circuit and contacts 1 and 3 in the high voltage circuit.

Actually, the Changeover Relay is a multipurpose DPDT relay with two sets of NO and NC contacts. In this application, only the NO contacts, 4 and 6 in the low voltage circuit and 1 and 3 in the line voltage circuit, are used. The Changeover Relay coil, wired to P_1 and C, is in the low voltage circuit.

With CR contacts 4 and 6 closed, current will flow from terminal R through the Compressor Contactor (CC, coil and the NC High Pressure Control (HPC) to Common (C).

When the Compressor Contactor coil is energized, it closes the NO Contacts (C and C) in the line voltage circuit and completes five circuits:

1. to the Potential Relay (PR) through terminals 5,2,1 and SC
2. to the Compressor, through terminals C and R
3. to the Defrost Timer (DT) motor
4. through the NC Defrost Relay (DR), contacts 1 and 2, to the Outdoor Fan Motor (OFM).
5. to the Reversing Valve Solenoid (RVS), through the now closed contacts, 1 and 3, of the Changeover Relay (CR).

When energized, the Reversing Valve Solenoid (RVS) will shift the valve to the heating mode, making the outdoor coil the evaporator and the indoor coil the condenser.

If the fan is set on *auto*, the Indoor Blower Motor (IBM), Figure 13-26, will be started by a fan control in the furnace, just as with other types of furnaces.

With the compressor running, the system is now in the heating mode.

When the compressor alone can no longer supply the necessary heat, the second-stage heating bulb in the thermostat will complete a circuit to terminal W_2, through the first Time Delay Heater (TDH1) to Common (C), Figure 13-24.

After a few seconds delay, the heater will close Main Contact 1, (MC_1) Figure 13-26, completing a circuit through Resistance Heater 1 (RH_1). At the same time, Auxiliary Contacts 1 (AUX_1). will close and complete a circuit from terminal R, through the second Time Delay Heater (TDH_2), to the first Outdoor Thermostat (OT_1).

If the outdoor temperature continues to fall, to the second balance point, the first Outdoor Thermostat (OT_1) will close and complete the circuit through the second Time Delay Heater (TDH_2). After a few seconds delay, the heater (TDH_2) will close a second set of Main Contacts (MC_2) and energize Resistance Heater 2 (RH_2). At the same time, the second set of Auxiliary Contacts (AUX_2) will close and complete a circuit from terminal R, through the third Time Delay Heater (TDH_3), to the second Outdoor Thermostat (OT_2). If the outdoor temperature continues to fall to the next balance point, the Outdoor Thermostat (OT_2) will close and complete a circuit through the third Time Delay Heater (TDH_3). After a few seconds delay, the heater (TDH_3) will close the third set of Main Contacts (MC_3) and energize the last of the three Resistance Heaters (RH_3).

The Outdoor Thermostats will remain closed as long as the outdoor temperature remains below their set points. Thus, the Resistance Heaters will cycle, as they do in an electric furnace, without having to wait for the outdoor thermostats to complete the circuit through the Time Delay Heaters.

3-BULB DEFROST

The defrost timer has a clock motor (DT) in the line voltage circuit that runs whenever the compressor is energized. After 90 minutes of operation (this can be shortened to 30 to 45 minutes) the clock will mechanically close a set of NO contacts between terminal 2 and the Defrost Relay (DR), Figure 13-25. The NO circuit, through terminal 1 and 2, is controlled by a temperature sensing bulb that must have good thermal contact with the coil, at a location that will be the last to reach the desired temperature. As frost builds up on the coil, its temperature will decrease and the bulb will close contacts 2-1. The closing point is adjustable and varies with the geographic area and design conditions. Most units are set between 30 and 35°F.

In order to defrost the outdoor coil, the Defrost Timer (DT) must have completed its cycle and the coil temperature must be below the set point of the Defrost Thermostat (DTh). Otherwise, the system will skip the defrost cycle for another 90 minutes. Therefore, the Defrost Thermostat (DTh) must close the circuit between contacts 1 and 2, Figure 13-25, and the Defrost Timer must have completed its cycle and closed the contacts between terminal 2 and the Defrost Relay (DR) coil.

When this circuit is completed by the combined action of the defrost system, the Defrost Relay (DR) coil is energized. This will open the line voltage contacts (DR) between terminals 1 and 2, opening the cir-

cuit to the Outdoor Fan Motor (OFM) and the Reversing Valve Solenoid (RVS) even though Changeover Relay (CR) contacts 1 and 3 are held closed by the Changeover Relay coil in the low voltage circuit. At the same time, the low voltage contacts, 4 and 6, of the Changeover Relay (CR) will maintain the circuit to the Compressor, to assure that it remains running. The system is now in the defrost mode and switched to the cooling cycle.

W_1 is no longer controlling heat from the compressor, therefore, the system must bypass the thermostat. When the Defrost Relay was energized, it also closed low voltage contacts 4 and 6 in the Defrost Relay (DR), completing a circuit from R, through the now closed Changeover Relay, to the first Time Delay Heater (TDH$_1$).

After a few seconds the TDH$_1$ and the first Resistance Heater (RH$_1$) will operate continuously during the defrost cycle. The other two Resistance Heaters (RH$_2$ and RH3) will be energized by the Outdoor Thermostats (OT$_1$ and OT$_2$), if needed.

When the coil temperature reaches the set point (about 65°) the Defrost Thermostat (DTh) opens and de-energizes the Defrost Relay (DR) coil. Or, the DR coil may be de-energized when the Defrost Timer motor completes its cycle, in about 10 minutes, and opens the Defrost Timer contacts (DT).

With the Defrost Relay coil deenergized, the Defrost Relay (DR) will restore the high voltage circuit to the Outdoor Fan Motor (OFM) and the Reversing Valve (RVS) through the still closed Changeover

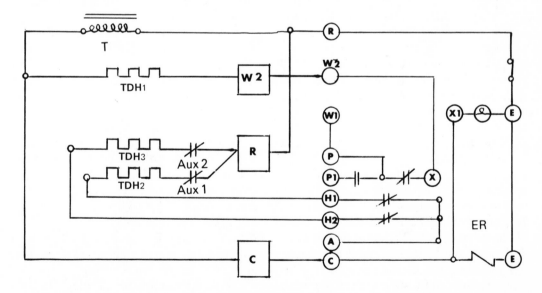

FIG. 13-28.   Low voltage circuits during emergency heat. Emergency heat switch is closed, the emergency heat relay is energized and the outdoor thermostats are by-passed.

Relay contacts 1 and 3. The system is once again in the heating mode.

At the same time, the low voltage contacts 4 and 6 of the Defrost Relay will open and restore system control through the thermostat.

3-BULB EMERGENCY HEAT

If the compressor is inoperative or down for service, heat may be maintained by manually closing the Emergency Heat Switch at the Thermostat, Figure 13-28. This completes a circuit to the signal lamp between terminals X_1 and E and energizes the Emergency heat Relay coil (ER) between terminals E and C. The ER relay then reverses the DPDT Switch between terminal P_1 and X, causing the NC contacts to open and the NO contacts to close. This opens the circuit to the Changeover Relay (CR) coil and links W_1 in series with W_2 through terminals P and X.

If the system is in the defrost mode, power will still be available to the compressor. Upon termination of the defrost cycle, contacts 4 and 6 of the Defrost Relay will open. Both the Changeover Relay and Defrost Relay are now open and there is no power to the compressor.

With W_1 calling for heat, W_2 will energize the first TDH and, in a few seconds, the first Resistance Heater.

When it was energized, the Emergency heat Relay coil (ER) closed the two sets of contacts from terminal A to terminals H_1 and H_2, thus bypassing the Outdoor Thermostats. The Resistance Heaters can now sequence like an electric furnace without being dependent on the Outdoor Thermostats.

4-BULB THERMOSTAT

A common 4-bulb wiring scheme uses a two-stage cool, two-stage heat thermostat with an emergency heat switch and light. There are a number of differences between these circuits and the ones shown previously. The major change is that the reversing valve solenoid is in the low voltage circuit.

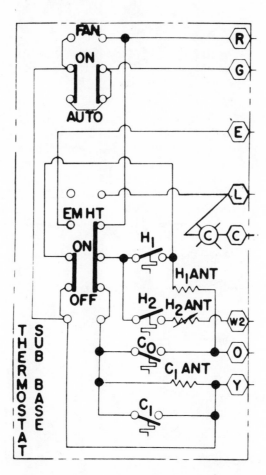

FIG. 13-29. Schematic wiring diagram of a four-bulb thermostat.

Note, in Figure 13-29, that the first stage heat anticipator is nonadjustable. Its resistance will reduce voltage below the pull-in voltage required by the reversing valve solenoid, so the reversing valve cannot be energized in the heating mode.

Note also that R, C, and W_2 go to both the indoor, Figure 13-33, and outdoor, Figure 13-31, units. G and E go only to the indoor unit and O and Y go only to the outdoor unit.

4-BULB COOLING

With the thermostat set to *on* and the

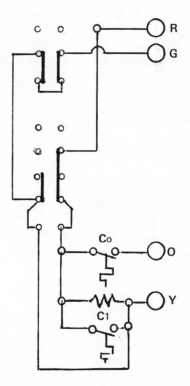

FIG. 13-30. Four-bulb thermostat in the cooling mode.

fan set to *auto*, the reversing valve is de-energized and in the heating mode. The crankcase heater (CH) and the line voltage side of the (T) were energized with the closing of the line voltage switch, Figure 13-32.

On a call for cooling, Figure 13-30, the first-stage, or Changeover bulb (Co), closes and energizes the Reversing Valve (RVS) Solenoid by completing a circuit from terminal R, through O, to the Reversing Valve in the outside low voltage section, Figure 13-31. With the Reversing Valve Solenoid energized, the system shifts to the cooling mode.

When the second-stage Cooling (C_1) bulb closes, it completes a circuit from terminal R, through terminal Y, to the Contactor Coil (CC), if the NC High Pressure Control (HPC) is closed. This pulls in the Contactors (C and C) completing circuits to the Potential Relay (PR), the Compressor (CSR), the Outdoor Fan Motor

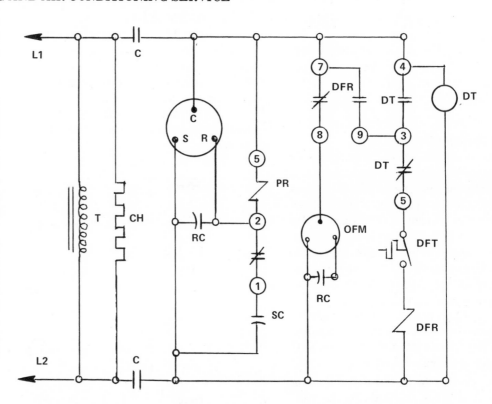

FIG. 13-32.   Outdoor line voltage circuits for a heat pump controlled by a four-bulb thermostat.

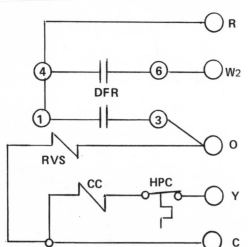

FIG. 13-31. Outdoor low voltage circuits for a heat pump controlled by a four-bulb thermostat.

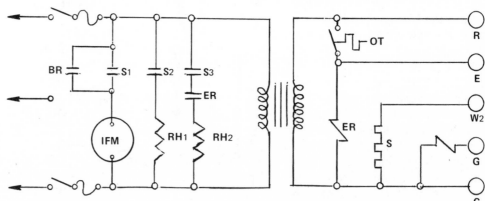

FIG. 13-33.   Indoor line voltage circuits for a heat pump controlled by a four-bulb thermostat.

(OFM) and the Defrost Timer motor (DT), Figure 13-32. A parallel circuit is completed, through the mode switch and fan control switch, to terminal G, to energize the Blower Relay (BR) in the indoor low voltage section. With relay energized, the Blower Relay contacts close and start the Indoor Fan Motor (IFM).

When cooling is no longer required, C_1 opens the circuit to the Blower Relay and to the Contactor Coil. This stops the Indoor Fan Motor, the Compressor, the Outdoor Fan Motor, and the Defrost Timer motor. When the temperature drops approximately 1° below the cooling set point, the first stage bulb (Co) opens to de-energize the Reversing Valve Solenoid and return the system to the heating mode.

4-BULB HEATING

On a call for heating, Figure 13-34, the first-stage heating bulb (H_1) closes and completes a circuit from terminal R, through the mode switch, to terminal Y. As with the cooling mode, this energizes the Contactor Coil and starts the Compressor, the Outdoor Fan Motor and the Defrost Timer (DT) motor. A parallel circuit is also established from H_1, through the mode switch, to terminal G, to energize the Blower Relay (BR) and start the Indoor Fan Motor (IFM).

When the second-stage Heating bulb (H_2) closes, it completes a circuit to W_2 and energizes the heater (S) in the Sequencer, Figure 13-33. The sequencer used in this example, Figure 13-35, has a heater in the low voltage circuit and three line voltage contacts, one for the blower and two for the Resistance Heaters (RH).

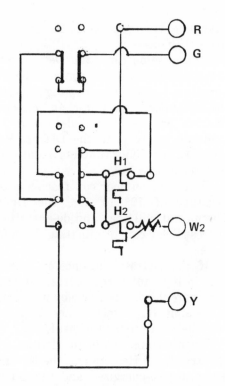

FIG. 13-34. Four-bulb thermostat in the heating mode.

Sequencing is by temperature differential. At its lowest design temperature, the Sequencer heater closes the contacts (S_1) to the Indoor Fan Motor. At a slightly higher temperature, it closes the contacts (S_2) to energize the **Resistance Heater** (RH₁). At a still higher temperature, it closes the **contacts** (S_3) to the second Resistance Heater (RH₂).

The second Resistance Heater (RH₂) is not energized until the Outdoor Thermostat (OT) closes and completes a circuit between terminal R and terminal E, Figure 13-33. This energizes the Emergency

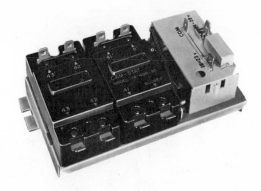

FIG. 13-35. Fan and sequence control. *Courtesy Cam-Stat, Inc.*

Relay (ER) coil and closes the second set of contacts (ER) in the RH₂ circuit.

While this example shows only two resistance heaters, there can be any number. Each new pair of heaters is controlled by an additional sequencer.

On an indoor temperature rise, the second-stage bulb (H_2) opens and interrupts the circuit to the Sequencer heater (S). As a result, the Sequencer contacts (S_1, S_2, S_3) open the circuits to the Resistance Heaters, even if the outdoor temperature has not risen above the Outdoor Thermostat setting.

On a further rise of the indoor temperature, the first-stage bulb (H_1) opens and shuts off the Compressor, Outdoor Fan Motor and the Defrost Timer motor.

4-BULB DEFROST

After the compressor has been running for 90 minutes, the Defrost Timer (DT) closes its NO contacts, between terminals 4 and 3, for 10 seconds, Figure 13-32. If the

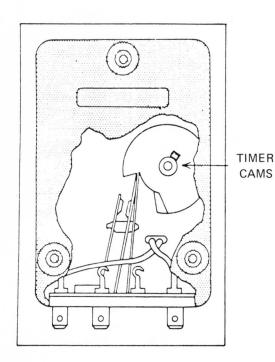

FIG. 13-36. Cutaway view of defrost timer showing timer cams.

Defrost Thermostat (DTh) is closed (at about 31°F) the Defrost Relay (DR) coil will be energized through the NC contacts 3 and 5. If the Defrost Thermostat is open, contacts 3 and 4 will be reopened in 10 seconds by the Defrost Timer cam and defrost will be deferred for another 90 minutes, Figure 13-36.

With the Defrost Thermostat closed, the Defrost Relay will close two sets of contacts, 4-6 and 1-3, in the low voltage circuit. This will assure a circuit from terminal R to terminal W_2, to sequence the Resistance Heaters, and another from terminal R to terminal O, to energize the

Reversing Valve and shift the system to the cooling mode.

At the same time, the Defrost Relay opens the line voltage contacts between terminals 7 and 8, interrupting the circuit to the Outdoor Fan Motor, and closes the contacts between terminals 7 and 9, to maintain a circuit to the Defrost Relay after Defrost Timer contacts 3 and 4 have been reopened.

When the outdoor coil reaches a temperature of 70° to 80°, the Defrost Thermostat will open and de-energize the Defrost Relay coil, terminating the defrost cycle.

If the Defrost Thermostat does not open in 10 minutes, the Defrost Timer will terminate the defrost cycle by momentarily opening NC contacts 3 and 5, to interrupt the circuit to the Defrost Relay.

Upon termination of the defrost cycle, the Resistance Heaters and the Reversing Valve are de-energized and the Outdoor Fan Motor is restarted when the two sets of low voltage contacts and the two sets of high voltage contacts in the Defrost Relay return to their normal conditions.

EMERGENCY HEAT

Emergency Heat is manually selected by moving the mode switch at the thermostat, Figure 13-37, to open the thermostat circuits to the Indoor Fan Motor, the Compressor and the Reversing Valve. Since the G leg is de-energized, the furnace fan control is now used to operate the Indoor Fan Motor.

At the same time, a circuit is closed between terminals L and C to energize the indicator light.

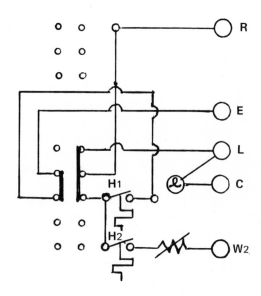

FIG. 13-37. Four-bulb thermostat in the emergency heat mode.

On a fall in temperature, the first stage heat bulb (H_1) closes and completes a circuit from terminal R, through the switch, to terminal E. The circuit from terminal E bypasses the Outdoor Thermostat, Figure 13-33, to energize the Emergency heat Relay (ER) and close the ER contacts in the second Resistance Heat (RH_2) circuit.

On a further fall in temperature, the second stage heating bulb (H_2) will close and complete a circuit, through W_2, to the Sequencer (S) heater. The Sequencer will, in turn, complete the circuits to the Indoor Fan Motor (S_1), the first Resistance Heater (S_2) and the second Resistance Heater (S_3).

Thereafter, the Resistance Heaters will be cycled by the second-stage bulb (H_2) alone.

Chapter 14
HEAT PUMP SERVICE

HEAT PUMP SERVICE

This section assumes that the system has performed satisfactorily in the past and now has a service problem. The previous sections described what the various components *should do* and these should be reviewed before attempting service on a unit. The charts shown at the end of the section can be a guide to your procedures but it is necessary to *think the problem out* before proceeding with a correction.

Air flow is very critical to a heat pump, so the filters must be clean, the duct system adequate and the temperature rise (or fall) across the coil must be correct for the unit to operate properly. Check these factors first on any service call, assuming the unit is operating.

Most trouble shows up initially as an electrical problem, so this is where any analysis should start.

INDOOR AND OUTDOOR DISCONNECT SWITCHES
1. Check for supply voltage to each disconnect — 240v.
2. Check for blown fuses. *Do not* assume that replacing a blown fuse will solve the problem. A short in the system blew the fuse. The cause must be found and repaired before applying power.

TRANSFORMERS
1. Check input voltage — 240v.
2. Check secondary voltage — 24v.
3. If either voltage is not correct, disconnect all wires, turn off power and check for shorts (infinite continuity) and resistance coil.

RELAYS
1. Check coil for voltage input. *CAUTION: Some will be 240v.*
2. Check across each contact for voltage. A voltmeter reading means the contact is open, no voltage means it is closed.
3. Disconnect all wires, turn off power and check for shorts. Check coil for resistance.

NOTE: Some wiring schemes will not have a connection to one of the DPDT terminals, so these cannot be checked by voltage. Look at the wiring diagram and check for continuity.

.OUTDOOR THERMOSTAT
1. Turn off power.
2. Check continuity across contacts — should be open if ambient is above dial setting.
3. Turn dial below ambient setting. Check for continuity across contacts. Should now be closed. If not, replace.

INDOOR THERMOSTAT
1. Thermostat should be checked for level.
2. With power off and thermostat not calling for heating/cooling, check each circuit for continuity — all should be open.
3. With power on, check the heating and cooling circuits for voltage. No voltage indicates circuit is closed, reading 24v means the circuit is open.
4. Check heat anticipation on both first and second-stage heating and set to the readings recorded. Note that *some* first-stage anticipators and *all* cooling anticipators are fixed.

SEQUENCER (WHITE-RODGERS)
1. Disconnect wires and check heaters for resistance (about 150 ohms plus or minus 20%).
2. Check contacts for continuity — both sets should be open.
3. Reconnect wires, turn on power, set thermostat to call for heat.
4. Check contacts for voltage. No voltage means they have closed.

REVERSING VALVES

Frequently, the reversing valve is blamed for many heat pump problems when it is really not at fault. It is a most important component but, before using the Ranco Touch Test, check the following:

1. Check the compressor discharge pressure. Low pressure will not operate the reversing valve.

2. Check the compressor suction pressure. High suction pressure indicates a leaking check valve.

3. Leak test and recharge, if needed. Normal suction pressure and low discharge pressure may indicate a defective compressor.

4. Energize the reversing valve. Remove the lock nut to free the valve's solenoid coil. Slide the coil partly off the stem to see if there is a magnetic force trying to hold it on.

CAUTION: Some coils are 240v.

When the coil is moved farther, a clicking noise will indicate the plunger has returned to the non-energized position. Replacing the coil on the stem should result in another click, indicating that the valve has responded by changing over.

5. Check for physical damage to the valve. Inspect for dents, deep scratches or cracks. When the above checks have been made satisfactorily, then perform the following Ranco Touch Test.

TOUCH TEST CHART

VALVE OPERATING CONDITION	DISCHARGE TUBE from Compressor	SUCTION TUBE to Compressor	Tube to INSIDE COIL	Tube to OUTSIDE COIL	LEFT Pilot Back Capillary Tube	RIGHT Pilot Front Capillary Tube	NOTES: *Temperature of Valve Body. **Warmer than Valve Body.	
	1	2	3	4	5	6	Possible Causes	Corrections
NORMAL OPERATION OF VALVE								
Normal COOLING	Hot	Cool	Cool, as (2)	Hot, as (1)	*TVB	*TVB		
Normal HEATING	Hot	Cool	Hot, as (1)	Cool, as (2)	*TVB	*TVB		
MALFUNCTION OF VALVE								
Valve will **not** shift from cool to heat	Check electrical circuit and coil						No voltage to coil.	Repair electrical circuit.
							Defective coil.	Replace coil.
	Check refrigeration charge						Low charge.	Repair leak, recharge system.
							Pressure differential too high.	Recheck system.
	Hot	Cool	Cool, as (2)	Hot, as (1)	*TVB	Hot	Pilot valve okay. Dirt in one bleeder hole.	Deenergize solenoid, raise head pressure, reenergize solenoid to break dirt loose. If unsuccessful, remove valve, wash out. Check on air before installing. If no movement, replace valve, add strainer to discharge tube, mount valve horizontally.
							Piston cup leak.	Stop unit. After pressures equalize, restart with solenoid energized. If valve shifts, reattempt with compressor running. If still no shift, replace valve.
	Hot	Cool	Cool, as (2)	Hot, as (1)	*TVB	*TVB	Clogged pilot tubes.	Raise head pressure, operate solenoid to free. If still no shift, replace valve.
	Hot	Cool	Cool, as (2)	Hot, as (1)	Hot	Hot	Both ports of pilot open. (Back seat port did not close.)	Raise head pressure, operate solenoid to free partially clogged port. If still no shift, replace valve.
	Warm	Cool	Cool, as (2)	Warm as (1)	*TVB	Warm	**Defective Compressor.**	
Start to shift but **does not complete reversal**	Hot	Warm	Warm	Hot	*TVB	Hot	Not enough pressure differential at start of stroke or not enough flow to maintain pressure differential.	Check unit for correct operating pressures and charge. Raise head pressure. If no shift, use valve with smaller ports.
							Body damage.	Replace Valve.
	Hot	Warm	Warm	Hot	Hot	Hot	Both ports of Pilot open.	Raise head pressure, operate solenoid. If no shift, replace valve.

VALVE OPERATING CONDITION	DISCHARGE TUBE from Compressor	SUCTION TUBE to Compressor	Tube to INSIDE COIL	Tube to OUTSIDE COIL	LEFT Pilot Back Capillary Tube	RIGHT Pilot Front Capillary Tube	NOTES: *Temperature of Valve Body. **Warmer than Valve Body.	
	1	2	3	4	5	6	Possible Causes	Corrections
Start to shift but **does not complete reversal**	Hot	Hot	Hot	Hot	*TVB	Hot	Body damage.	Replace valve.
							Valve hung up at mid-stroke. Pumping volume of compressor not sufficient to maintain reversal.	Raise head pressure, operate solenoid. If no shift, use valve with smaller ports.
	Hot	Hot	Hot	Hot	Hot	Hot	Both ports of Pilot open.	Raise head pressure, operate solenoid. If no shift, replace valve.
Apparent **leak** in heating	Hot	Cool	Hot, as (1)	Cool, as (2)	*TVB	**WVB	Piston needle on end of slide leaking.	Operate valve several times then recheck. If excessive leak, replace valve.
	Hot	Cool	Hot, as (1)	Cool, as (2)	**WVB	**WVB	Pilot needle and piston needle leaking.	Operate valve several times then recheck. If excessive leak, replace valve.
Will **not** shift from heat to cool	Hot	Cool	Hot, as (1)	Cool, as (2)	*TVB	*TVB	Pressure differential too high.	Stop unit. Will reverse during equalization period. Recheck system.
							Clogged Pilot tube.	Raise head pressure, operate solenoid to free dirt. If still no shift, replace valve.
	Hot	Cool	Hot, as (1)	Cool, as (2)	Hot	*TVB	Dirt in bleeder hole.	Raise head pressure, operate solenoid. Remove valve and wash out. Check on air before reinstalling, if no movement, replace valve. Add strainer to discharge tube. Mount valve horizontally.
	Hot	Cool	Hot, as (1)	Cool, as (2)	Hot	*TVB	Piston cup leak.	Stop unit, after pressures equalize, restart with solenoid de-energized. If valve shifts, reattempt with compressor running. If it still will not reverse while running, replace valve.
	Hot	Cool	Hot, as (1)	Cool, as (2)	Hot	Hot	Defective Pilot.	Replace Valve.
	Warm	Cool	Warm, as (1)	Cool, as (2)	Warm	*TVB	**Defective compressor.**	

VALVE OPERATED SATISFACTORILY PRIOR TO COMPRESSOR MOTOR BURN OUT—caused by dirt and small greasy particles inside the valve. **To CORRECT:** Remove valve, thoroughly wash it out. Check on air before reinstalling, or replace valve. Add strainer and filter-dryer to discharge tube between valve and compressor.

DEFROST TIMER

If the system is not defrosting, or defrosting erratically, check the defrost timer. No defrost cycle suggests a faulty timer motor or open contacts. Erratic defrost suggests that contacts are stuck closed. Check as follows:

CAUTION: All points are 240v.

1. Check power to the timer motor with compressor running.
2. Observe whether or not cam is turning, while compressor is operating, and proper voltage is supplied to the timer motor. If not, disassemble and check cam set screw for tightness.
3. Place a jumper wire across the defrost control contacts.
4. If system changes to defrost, remove wire from terminal 7 of the defrost timer, Figure 13-32. If system continues in defrost cycle, timer contacts 4-3 are stuck in the closed position. Wait 10 minutes to see if contacts 3-5 open and terminate cycle. Turn off power. Remove wires and check terminal 4-3 for continuity. If closed, replace timer.
5. If the unit has a *heavily* frosted coil, jumper terminals 4-3. If the unit goes into defrost, the contacts are open. Recheck Steps 1 and 2. If these are correct, replace timer.

DEFROST CONTROL

The defrost control normally has a knob to adjust the point at which defrost is initiated. This can be anywhere from 30° to 45°F. It will terminate defrost at 65° to 75°F, according to the manufacturer's design. Note the set points before proceeding. Be sure the sensing bulb is making good contact with the coil. If in a well, make sure there is enough thermo-mastic to make good contact and that the capillary is inserted completely.

1. An accurate temperature sensor must be used to exactly determine when defrost is initiated and terminated.
2. Check the temperature when the defrost cycle is initiated and terminated. It should be within 4° to 5°F of the manufacturer's settings. If not, replace.
3. If no defrost is initiated when the coil temperature falls below the settings, install a jumper between terminals 3 and 4 of the defrost timer. A defrost cycle should be initiated. If not, place a second jumper across the defrost control. If this starts a defrost cycle, the control should be replaced.

CHECK VALVES

The usual check valve has a steel ball bearing that opens under pressure from one direction and seats closed under pressure from the opposite direction. Its operation can be checked in two ways:

1. With the unit off, the ball bearing can be cycled with a large magnet. Listen for the sound of the ball seating.
2. With the unit operating, there will be a definite temperature difference across the valve if it is closed and no temperature difference if it is wide open. This can be determined by touch.
3. If the valve is leaking, the temperature difference will be much less than normal. Checking for this problem requires good measuring devices.

CHARGING HEAT PUMPS

Having the proper refrigerant charge in a heat pump is a major factor in its efficient operation. New units come with the proper charge and have precharged line sets that are connected in essentially the same way as air conditioning line sets. There should be no question of proper charge when connecting the heat pump for the first time, either summer or winter. On a service call, the situation may be more difficult. The unit may have lost charge or it may be necessary to remove the charge before servicing.

The most accurate and efficient way to recharge the system is to weigh the refrigerant, following the manufacturer's instructions and specifications. This method will assure a correct charge, regardless of the ambient temperature.

As often happens, the manufacturer's specifications are not available, or the system may need only a partial charge. In these cases, other charging methods must be used.

Several manufacturers publish charging charts that relate suction and head pressure to the dry bulb and wet bulb temperatures of the coils, Figures 14-1 and 14-2. Separate charts are published for charging in the heating mode and in the cooling mode.

To charge in the cooling mode, measure the dry bulb temperature at the outside

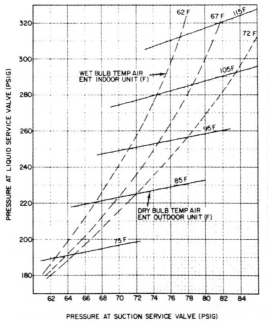

FIG. 14-1. Charging curves for high ambients.

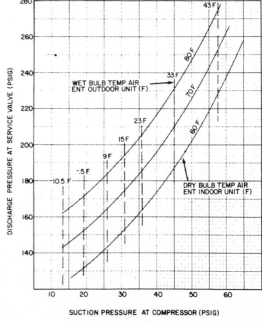

FIG. 14-2. Charging curves for low ambients.

coil and the wet bulb temperature at the inside coil. Assume that you read an outdoor ambient of 85° db and an indoor ambient of 67° wb. Begin charging the unit, as you would a single-purpose air conditioner. Both the head and suction pressures will rise as you add refrigerant. From the chart, you can see that the unit is properly charged when the suction pressure is 70 psig and the head pressure is 224 psig. When the heat pump is properly charged in the cooling mode, it will be properly charged in the heating mode.

It is very easy to overcharge when the ambient is below 64° to 75°, so care should be taken below these temperatures. Remember, too, that head and suction pressures will generally run 7 to 8% lower on a heat pump than they will on a straight air conditioner with R-22.

To charge in the heating mode, measure the wet bulb temperature at the outside coil and the dry bulb temperature at the inside boil. If you measure 33° wet bulb and 70° db, Figure 14-2, the unit will be fully charged when the head pressure is 210 psig and the suction pressure is 45 psig.

Another method of charging, in the cooling mode only, is by superheat. Repeated temperature and pressure measurements must be taken when using this method.

First, measure the suction pressure and temperature at a point very close to the compressor. Assume they are 60 psig and 70°F. Reading from Table 8-1, the saturated or unsuperheated vapor should be 34°. Therefore, the superheat is 36°, which is excessive.

From the chart below, select the superheat required at the working ambient.

Ambient db	Superheat
100° - 110°F	10°
90° - 99°F	15°
70° - 89°F	20°

If the ambient measures 85°, the superheat should be 20° and the suction vapor temperature, at 60 psig, should read 54°, not 70°. However, the suction pressure will rise as the refrigerant is added to reduce the superheat, so this charging method is one of trial and error.

Say, on the next reading, that suction pressure is 70 psig and suction temperature is 65°. Saturated temperature at this pressure is 41°, so superheat is now 24°.

Adding more refrigerant raises the suction pressure to 73 psig and lowers the suction temperature to 63°. Saturated temperature at this pressure is 43°. Therefore, the system is operating with 20° superheat and is properly charged.

The above method *will not work* if the unit must be charged below 65°F. The manufacturer's published charging methods *must be* used when charging in low ambients.

OTHER COMPONENTS

Components common to air conditioning; compressors, capacitors, blower motors, contactors, etc., plus the electric heat elements with their limits and fusible links are checked in the manner described in earlier chapters.

OVERALL TROUBLESHOOTING

The following charts give a logical troubleshooting sequence for both the heating and cooling modes. These do not have to be memorized since most top manufacturers issue similar materials to their service departments. Knowing the operation of the various controls and how they fit into the overall system are the most important service skills.

TROUBLESHOOTING CHART — COOLING CYCLE

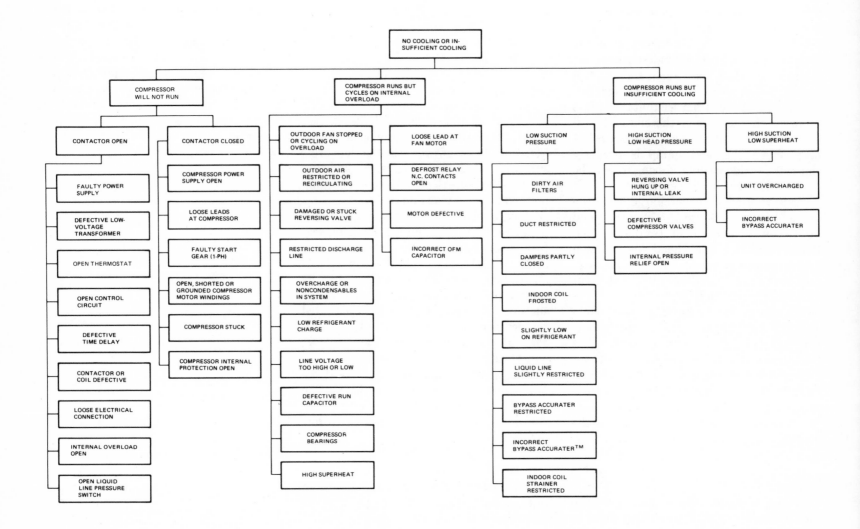

Courtesy Carrier Corp.

TROUBLESHOOTING CHART — HEATING CYCLE

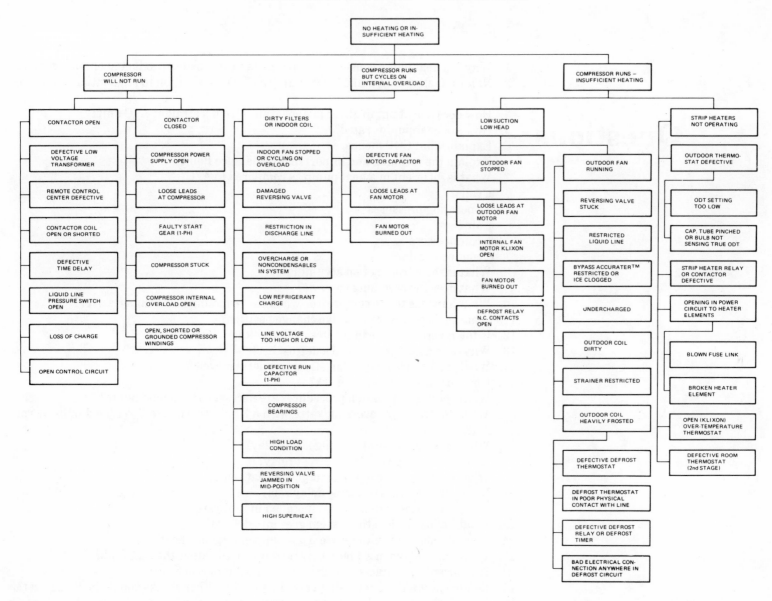

Questions

1. Why is a heat pump considered to be the fastest growing product in climate control?
2. Name 5 factors that make a heat pump a practical solution to residential climate control.
3. How does a heat pump differ in appearance from a straight A/C unit?
4. How are heat pumps rated?
5. Explain what COP means.
6. Explain the meaning and importance of Balance Point.
7. What is different in a heat pump as compared to A/C?
   a. coils
   b. operating pressures
   c. air flow
   d. compressors
   e. connecting lines
8. What is the purpose of an accumulator?
9. What does the reversing valve do?
10. What happens to the compressor when the reversing valve changes over?
11. What makes a reversing valve change position?
12. Is the reversing valve in the low or line voltage circuit?
13. Why does a heat pump need a defrost cycle?
14. What factors are involved in starting a defrost cycle?
15. How does a defrost timer work?
16. What special installation procedures are necessary for a heat pump?
17. What controls the reversing valve with a 3 bulb thermostat? With a 4 bulb thermostat?
18. What is the purpose of a changeover relay?
19. What is the purpose of the defrost relay?
20. What is the purpose of the emergency heat relay?
21. Name the elements in a time delay relay.
22. Why is an outdoor thermostat installed in the system?
23. What happens when the system goes into defrost?
24. What happens when the system goes into emergency heat?
25. What changes when a 4 bulb thermostat is used rather than a 3 bulb?
26. How are relay contacts checked when the system is operating?
27. Before checking the reversing valve itself, what other checks should be made first?
28. What is the effect of low charge on the reversing valve?
29. How is a heat pump charged in the cooling mode? In the heating mode?

Chapter 15
COMMUNICATIONS

Communications covers many broad areas. It includes written and visual communications in addition to verbal communications. In talking to a customer, both what is said and what is not said are important. The serviceman accepts a dual responsibility, one to his employer and the second to the customer. The way in which he responds to these dual obligations will, to a great degree, determine his value to each.

The serviceman is often the only direct person-to-person contact that the customer will have with his employer. Therefore, whatever impression he leaves with the customer will be applied to the firm as a whole. The serviceman must not only be a specialist in heating and air conditioning service, he must also be a goodwill ambassador for his employer, for his actions on the job will determine whether the customer will deal with his employer in the future. Thus, the better the employee can develop his communications skills, the more valuable he will be to his employer, and the easier it will be for him to do his job efficiently and satisfactorily.

EMPLOYER RELATIONS

The serviceman has a number of fundamental responsibilities to his employer which he should always keep in mind when he is on the job.

1. *Appearance.* A serviceman's appearance should immediately give the impression that he is competent and knows what he is doing. He should be clean-shaven, neat and have a fresh uniform. He should approach each job in a very businesslike manner so that the customer will have confidence that he knows what he is doing.

2. *Care of Equipment.* Since the employer has spent a considerable amount of money in supplying the serviceman with a truck, tools and service instruments, the serviceman should treat them as his own and take good care of them. Driving the truck like a hot rod creates a bad impression on any observer and, since the employer's name is printed on the truck, it creates a bad impression for the employer. Many of the test instruments that he has provided are delicate instruments and must be cared for so that they will remain operative and useful for an extended period of time. Taking care of tools and instruments shows a basic respect for other people's property, which is a trait of all valuable employees. All tools and instruments should be stored in an orderly fashion and the instructions, provided with the instruments, should be followed exactly to prevent damage.

3. *Honesty.* The serviceman must be honest, both in what he says to the customer and how he reports back to the employer.

It is an easy temptation to *borrow* some of the parts or supplies stored in the truck for his own use, but fundamentally this is stealing and costs his employer money to replace those items. *Borrowing* has the same effect as *theft* by a third party.

4. *Safety.* The employee must follow proper safety precautions in using all of the tools and equipment provided to him, not only to prevent harm to himself but also to avoid an accident in the customer's home. The employer is liable for any damage by the employee on the job, or going to and from the job, and so the serviceman should be conscious of the safety factor at all times.

5. *Accuracy.* Accuracy includes doing the job right the first time *and* examining *all* of the factors that may have caused the original complaint. It is sometimes easy to locate and replace a component which has failed, and temporarily have the system back in operation, when in actuality the true cause of failure has

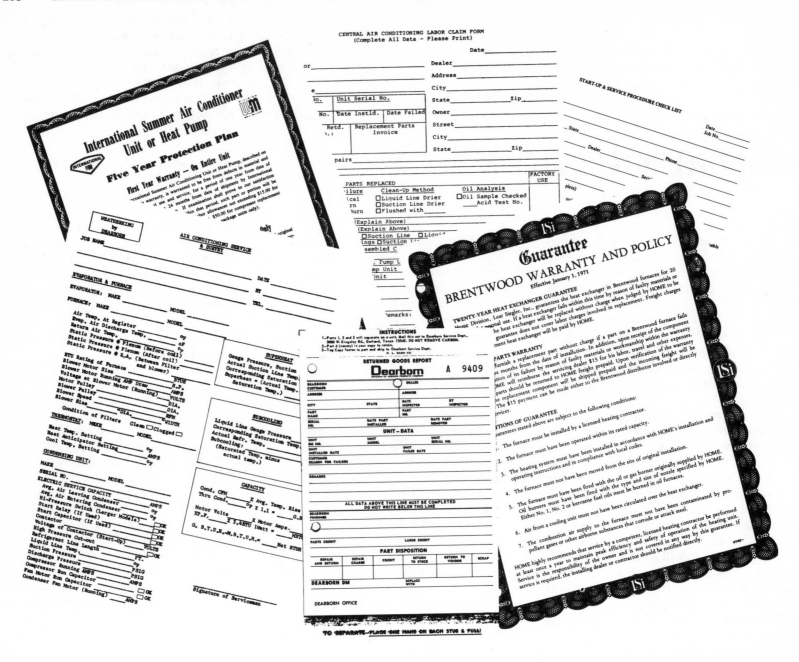

not been determined. As a result, the system will fail again. Call-backs, for incomplete or inaccurate jobs, are expensive to the employer. They also diminish the confidence the customer has in you, since the call-back increases charges against the job or to the customer.

6. *Following the Schedule.* Most servicemen are scheduled for a specific number of calls. As a result, the customer usually expects him by a certain time. While an extra cup of coffee is always enjoyable, it does reduce the number of productive hours available to the serviceman. And, if he is late arriving at the customer's home, the customer may be impatient and upset.

WRITTEN COMMUNICATIONS

Most organizations require written reports at the end of each day or at the end of each week, so that they can keep track of the activities of a given serviceman and evaluate how he is performing on the job. A few of these reports are listed to give an idea of the form they might take.

1. *Inventory Reports.* An inventory of spare parts and supplies must be carried on the truck and is checked out daily or weekly from the main office. As these parts are used during his service calls, the inventory report should be used to record which job required the parts and the other supplies that were used. This inventory control assures the serviceman that he will have an adequate supply of parts and supplies available on the truck when needed, rather than having to return to the main office. Loss or breakage of the tools and instruments which are assigned to the truck should also be reported, along with a clear explanation of how this happened.

2. *Cash and/or Check Accounting.* Many service firms attempt to collect immediately for the service call by having the serviceman present a bill to the customer. To do this, the serviceman must know the proper charges, including his chargeable labor rate, and the cost to the customer for materials used. He must fill out the bill completely, total it accurately and present it to the customer. The customer may pay for the call either in cash or by check. The amounts collected must be reported back to the main office and tied to a specific job so the customer can be given proper credit.

3. *Day's Call Reports* (sometimes called surveys). Many organizations require their servicemen to furnish additional information on each call so that they can keep adequate service records. This report lists the materials used, the cause of the service complaint and what was done to repair it. The inventory and cash accounting reports, previously mentioned, also become part of that customer's permanent record so that, at any future time, the dealer will know what was done on the previous call.

The serviceman should have his eye open for possible future sales of add-on equipment, such as air conditioning, humidifiers or electronic air cleaners. This type of information, along with the age and make of the customer's furnace, will help the dealer in planning a future sales campaign for additional add-on business. All of this information becomes part of the employee's and customer's permanent record. If the dealer is planning a spring promotion of add-on air conditioning, his service records pinpoint the customers who do not have air conditioning. These names become his major sales leads for either a mail or direct solicitation. Frequently, the serviceman is given a bonus or flat fee when he assists in making future sales.

4. *Return Goods Tag.* Any part that is replaced, while still in warranty, is returned to the dealer so that he can obtain proper credit from the manufacturer. The tag has several parts: one is a cardboard section that goes with the part and others are for dealer and manufacturer's records. The Return Goods Tag shows the date and place of installation, date replaced, part type and serial number, reason for failure and other information which will allow the manufacturer to easily identify the problem. Some manufacturers have a labor allowance, and will pay for a portion of the labor required to change the part. This usually applies to larger components. In this case, labor time must also be recorded. The Return Goods Tag will have a wire or other method for attaching it to the defective part so that the two will not become separated in transit to the original manufacturer.

The serviceman must be alert to all sales possibilities, particularly replacement sales. If the customer has a particularly old furnace or one which requires some major service, he might approach the homeowner along these lines: "Mrs. Brown, we can get your furnace operating for you. But, because of its age, there is a good chance of another breakdown in the near future. This might be inconvenient for you in the middle of winter. We can replace the system now, in less than a day, without any discomfort or inconvenience to you. The additional cost for a new furnace will assure you many more years of satisfactory service and comfort."

If the heat exchanger is cracked or badly rusted, point out the safety hazard to the customer. However, *be sure what you say is true.* Do not oversell! You might show her the problem and then say, "Mrs. Brown, see these cracks (or badly rusted spots)? If these break through, you may get an odor in the house and perhaps some carbon monoxide. We could replace this heat exchanger, or get you a new furnace next week, so that this will not be a future problem for you."

Customers will tend to have greater confidence in a serviceman pointing out these things than in a salesman. The serviceman benefits the customer by preventing future problems and inconvenience, and he builds sales volume for his employer and makes more money himself.

RADIO USAGE

Many modern service trucks are equipped with a two-way CB (Citizen Band) radio for constant communication with the main office. Additional routing is made via the radio, and emergency calls are relayed to servicemen in the field. The radio is a very convenient link to the main office. It allows the serviceman to utilize his time on the job rather than in reporting back to the office each day or in telephoning after each job is completed. Radio helps the home office keep track of the serviceman and to know when he has completed each job and is on to the next. It should not be used for personal business nor for communications that do not have a direct bearing on the current job. There are also some basic rules, established by law, covering the use of the CB radio. instructions come with the radio and should be thoroughly understood by the serviceman before using the CB radio.

CUSTOMER RELATIONS

Customers come in all sizes, shapes and kinds and will have the same interests, hang-ups and problems as you do. In the final analysis, they are merely people, just like you, who need recognition, assurance and confidence that you will take care of their immediate problem. The basic rule in all customer relations was written down in the Bible and states, very simply, "Do unto others as you would have others do unto you." Thus, if you treat someone else courteously and with an interest in their problem, the odds are that they will react to this in a favorable way and treat you in a similar manner.

Remember that the customer has paid good money for the heating and air conditioning system in his home and expects it to operate properly. If it does not, he expects the trouble to be corrected promptly and economically. He also did not buy the furnace or air conditioner because of its pretty color, clean design or the number of Btu's it will produce. Rather, he bought *comfort* in the living space, plus *convenience* and *dependability* of operation. Further, he must have confidence in you as the serviceman, and your ability to correct his problem. At the time of the service call, he may be frightened by a safety hazard; he may be mad because the furnace or air conditioner, for which he had just paid a considerable amount of money, does not function properly; he may be most uncomfortable (either too hot or too cold) because of the failure; or they may be all of these things. The serviceman's job is to establish himself as the expert who can take care of these problems and get the system back in operation in minimum time and lowest cost to the customer.

NEW INSTALLATION

After completing a new installation, it is a wise policy and good customer relations to show the customer how to get maximum comfort, with the least effort, from the new equipment.

Here are some comfort tips for your customer:

1. Set thermostat at comfort level (about 73° F) and leave it alone all year.
2. Be sure registers and returns are not blocked by furniture or drapes.
3. Keep doors and windows closed.
4. Use kitchen exhaust when cooking.
5. Have clothes dryer vented.
6. Change filters regularly.

On initial installations, the warranty provisions should be explained to the owner so he will know what is covered and what is not. Warranties will vary with the manufacturer and type of component, but some common warranty periods are:

Parts and most components—1 year if defective in manufacture.

Compressors—1 year with an optional 4-year extended warranty at extra cost.

Heat exchangers—5 to 10 years depending upon place of installation, etc.

Explain how to call for service if anything goes wrong. If your shop offers a service contract for regular service checkups, this may be explained to the homeowner at this time.

THE SERVICE CALL

The first thing to do, before making the service call, is to double-check your own appearance and grooming. Be doubly certain that your shoes are free of mud or dirt. Normally, the first step in the service procedure will be to check the thermostat. This is usually located in the living room and, more often than not, this area is carpeted. Tracking dirt across the customer's carpet is the surest, quickest way to upset her and have her lose confidence in your ability to do the job.

In your initial contact, time will be very well spent in asking her a series of direct questions in order to find out just exactly what seems to be wrong and why she made the service call. She may not be at all knowledgeable as to the basic symptoms, but she can answer general questions with sufficient clarity to give you a pretty good idea of where the trouble lies.

The newspaperman's habit of asking Who, What, Where, When and Why will help you find out what you need to know to fix the problem. Certain questions will give the customer an opportunity to provide you with some necessary and pertinent information so that you will have a better chance of going directly to the cause of the trouble and correct it in the shortest possible time. Questions such as the following might be asked:

"Mrs. Brown, your call to the office said you were having trouble with the furnace. Can you tell me what kind of trouble you have experienced?"

"When did you first notice this problem?"

"Did it happen only once or has it happened several times lately?"

"Have you ever had this same kind of problem before?"

"Do you notice it on certain days or certain times of the day?"

Answers to these questions will give you the necessary information which can help you locate the trouble.

It may also be necessary to pursue an answer to one of these questions, since the initial answer may be somewhat vague. For instance, you may ask the question, "What seems to be the trouble?" and the customer replies that she is not comfortable. This vague response does not give you enough information to go on, so you ask a second question, "When you say you are not comfortable, do you mean that you are too hot or too cold?" She may then reply that she is too cold. This still needs more explanation, so your next question might be, "When does this happen?" Her response might be that she is cold in the morning, but later on in the day this problem seems to clear up. By asking additional questions, the serviceman accumulates useful information and thus will know the general area that needs his attention.

As soon as the serviceman has obtained as much information as possible, he should, at this point, assure the customer that he knows the general area of the problem and can fix it to her satisfaction. He should *not*, at this point, specifically pinpoint the problem, since it could be something else. If, at the close of the call, he tells her what he actually did, she may suspect him of charging her for more than what was necessary. Assuring the customer that he can take care of the problem is all that is important at this point.

Keep in mind, at all times, that the major concern of the housewife is to have the problem fixed and the equipment back in operation. She doesn't know why the problem occurred nor how to fix it—she is depending upon you to take care of these areas. Whatever the problem is, it probably has caused her some inconvenience and perhaps discomfort. She will be most pleased that somebody is there to solve it. If you can assure her that you understand the problem, and feel that you can fix it in a reasonable time, she will normally be satisfied.

Courtesy is always of prime importance when meeting someone new. Most housewives will be extremely flattered if you answer her questions with "yes ma'am" and "no ma'am", rather than just a straight yes or no.

Be sure to keep the conversation on the problem at hand rather than talking about the weather or what a pretty house she has. Most servicemen are paid by the hour so, while these pleasantries are nice, it is costing her money to pass the time of day. As stated earlier, a businesslike, courteous approach builds her confidence in your efficiency and assures her that you can fix the problem.

It is particularly irritating to a serviceman to have an emergency call when the problem is quite minor and could have been fixed by the homeowner. Getting out of bed at 2:00 AM and finding the thermostat not set correctly, a blown fuse, or just the pilot out, can tempt him to make some nasty remark about the customer's intelligence. While it may give some temporary personal satisfaction, it is never good business to make a customer look foolish. After all, he is paying for the call.

Occasions arise where the customer, for one reason or another, will be agitated or mad and attempt to blame you. In all cases of this nature, the fundamental and prime rule is, first, to hear her out completely. Listen carefully, paying attention to what she says. It may be that she is just trying to get it out of her system and will calm down once she has said it all. Listen, because some of the things she says may be quite valid and will give you an opportunity to explain the situation in more detail. She may not have all the information at hand nor understand the circumstances of the problem but, if you let her talk it out, she may answer her own objections.

For example, assume that a new furnace installation has failed, in 3 or 4 months, and is still under warranty. It is winter and the house is cold, so she is both uncomfortable and upset because the furnace has broken down. She certainly has every reason to be upset under these circumstances and your first and immediate action should be to apologize for her inconvenience and problem.

You might say: "We're certainly sorry, Mrs. Brown, that you have had this inconvenience. I know that it's most uncomfortable in here with the furnace being down. Because of this, we put your call on a priority basis. I got here just as soon as possible to get the furnace back in operation for you. You also understand that this is under warranty and this call will be at no expense to you. I hope to have the trouble corrected very shortly and we will get the house warmed up again in a short time."

Assume that you are several hours late in arriving at the customer's home, due to some unusual problems you experienced with the previous service call. The customer is upset and tells you that she's been waiting around all afternoon for you to come, which has been an inconvenience. In the first place, where possible, you should not let this happen. If you see you are going to be running late for your next call, it would be a courteous thing to ask for the use of a phone, or use the truck radio, to either call yourself or have the office call your next customer, explaining that you will be late. However, once this happens, you should not offer excuses. If there are some extenuating circumstances, they can be explained in a simple, direct way. Again, it is most important that first you apologize for the inconvenience and whatever problem you have caused, then assure her that this is not the way things usually happen.

Another customer may have smelled gas and become concerned that there could be a fire or explosion in the house. Many people can become extremely fearful with a gas or oil smell in the house. In this type of situation, you should immediately ascertain the cause of this problem, return and explain that there is no danger, even before you fix it. There will often be a slight gas smell if the pilot light has gone out. Many people are quite sensitive to gas and recognize it even in very minor, trace amounts. Again, your assurances that there is no great danger, and that you will correct the problem immediately, is your best course of action.

In some cases, you will find a customer who is highly critical of the equipment, the comfort level it maintains, or anything else pertaining to their heating and air conditioning system. If your assurances do not satisfy the customer, you can advise her that you will have the man who sold her the equipment, or the owner of the company, get in touch with her immediately to answer her objections. This type of customer may be working on some incorrect assumptions or be a chronic complainer. In either case, it is important that the salesman or owner have an opportunity to answer her objections since this kind of an attitude can result in lost sales, in the future, to all of her friends and acquaintances.

Another customer may be quite appreciative of your correcting the problem and be very friendly. This one might offer you coffee or iced tea after you have corrected

Communications   243

her problem. The serviceman must, in a very courteous way, decline this offer. He can explain that he is paid by the hour and his time taking a coffee break would be charged to the customer. He can also mention that he has several other calls to make and would not wish to keep other customers waiting. Most people would realize that this would be an inconvenience to someone else and will accept your refusal graciously.

Most servicemen replace the furnace filters as a normal part of the call, to make sure that they will function correctly in future months. It is usually appreciated if you suggest to the housewife that she watch how this is done, so that her husband can replace them when the filters become clogged again. Here is a good opportunity to explain the purpose of the furnace filters and why they need to be changed at regular intervals. It is important to do this in a positive way, pointing out the more efficient operation of the furnace rather than suggesting that the house or basement is dirty.

In a similar manner, give her a brief explanation of the thermostat and how she can obtain maximum comfort for the living space. Suggest that she turn on the blower control to constant operation to avoid stratification and unsatisfactory comfort levels. Also suggest that she pick a pleasing comfort level and leave the thermostat set at this point in both winter and summer. Suggestions like these create good customer relations and result in a more satisfied customer.

CLOSING THE CALL

In closing the call, it is of primary importance to assure the housewife that the problem is fixed and should not occur in the future. You should carefully explain exactly what was the problem, what was done to correct it and what she can now expect. Any parts that were needed should be itemized on her bill so that she will know the exact material costs. It is also good practice to leave the defective parts so that she and her husband can examine them themselves. (If the parts are in warranty, she is not charged and the defective parts are returned for dealer credit.)

As mentioned earlier, some companies attempt to get paid at the time the work is completed. If this is the case, the bill should be presented and each item explained in as much detail as she wishes. The main purpose of this explaination is to build her confidence that the problem has been corrected and that she will experience the comfort conditions she expects.

The serviceman should also make sure that the area where he was working is left clean and neat: all parts picked up, any boxes or wrappings collected, wire ends or tape removed, the floor swept and smudge marks or oil on the furnace wiped off. Any parts that have been replaced should be left in one place and the customer advised where they are. If the filters have been replaced, the old filter should be taken out and thrown away, as well as the other debris that has been collected.

Finally, he should leave the company name and address, plus telephone numbers for day and night service, with the customer. Many companies print stickers with this information, plus space for the date of the last service call. These stickers are attached to the furnace for ready reference.

Sometimes it is not possible to fix the problem. If this is the case, carefully explain to the homeowner why the problem cannot be fixed and what she must do next.

If the problem requires a special part, this should be explained to her and its availability checked out with the main office. If it has to be ordered, then she should be given a reasonable schedule as to when it might arrive and when you can return to make the replacement. If the part is in stock at the main office, she might be given a choice of your returning and repairing it the same day, or finishing the job on the following day. If the furnace doesn't run at all and the outside temperature is low, she may want you to return the same day.

A most unusual circumstance is where the electrical circuits are overloaded, requiring either rewiring to the junction box, or shifting some other appliances from the main supply voltage. This is normally a job for a qualified electrician and one that the residential heating and air conditioning man is not equipped or licensed to handle. Most dealers will have a working arrangement with a licensed electrician. The homeowner should be advised of the severity of the problem, with the suggestion that he call the original electrical contractor or his own electrician. If he has no preferences, you may recommend an electrician who has previously done some work with you. This informs the customer of the gravity of the problem and offers him a possible solution and an alternative, even though you were not able to correct the difficulty yourself.

SUMMARY

This section has attempted to point out just some of the basics of dealing with other people. It is not really necessary for the serviceman to understand the reasons why people act as they do, but rather to respond to their actions in a courteous and positive way. If the serviceman asks a number of questions at the beginning, so that he gets answers that help him solve the problem, then goes about solving the problem as quickly and efficiently as possible, he is going to have a satisfied customer. He should, at all times, be courteous and listen carefully and attentively to the customer's story, complaints or anything else that she wants to tell him, and then go about his business and get the job done. A smile goes a long way towards gaining a satisfied customer!

Finally, he has to explain to the customer what he has done and assure her that this will result in her getting the comfort and convenience that she expects and deserves. If he can gain her confidence, calm any fears that she may have, fix the problem and have everything working properly when he leaves, he has given proper service to his customer and helped the employer build future business.

Questions

1. Name 6 basic responsibilities to your employer.
2. Why are inventory reports important?
3. When is a return goods tag filled out? What is done with it?
4. Why have call reports or surveys?
5. What is a CB Radio?
6. Why have trucks equipped with CB radios?
7. What does a customer buy?
8. Name some things to tell the customer on a new installation.
9. What is the first thing to check on a service call? Why?
10. How do you get the information you need? Give some examples.
11. Should you tell the customer you know what's wrong? Why?
12. What is the basic rule in customer relations?

INDEX